Plants & Gardens

Brooklyn Botanic Garden Record

Soils

1990

Brooklyn Botanic Garden

STAFF FOR THE ORIGINAL EDITION:

JOHN PAUL BOWLES, GUEST EDITOR

BARBARA B. PESCH, EDITOR

WILLIAM C. MULLIGAN, ASSOCIATE EDITOR

STAFF FOR THE REVISED EDITION:

BARBARA B. PESCH, DIRECTOR OF PUBLICATIONS

JANET MARINELLI, ASSOCIATE EDITOR

AND THE EDITORIAL COMMITTEE OF THE BROOKLYN BOTANIC GARDEN

BEKKA LINDSTROM, ART DIRECTOR

DONALD E. MOORE, PRESIDENT, BROOKLYN BOTANIC GARDEN

ELIZABETH SCHOLTZ, DIRECTOR EMERITUS, BROOKLYN BOTANIC GARDEN

STEPHEN K-M. TIM, VICE PRESIDENT, SCIENCE & PUBLICATIONS

COVER PHOTOGRAPH BY ELVIN MCDONALD
ALL PHOTOGRAPHS BY ROBERT KOURICK, EXCEPT WHERE NOTED

Plants & Gardens, Brooklyn Botanic Garden Record (ISSN 0362-5850) is published quarterly at 1000 Washington Ave., Brooklyn, N.Y. 11225, by the **Brooklyn Botanic Garden, Inc.** Second-class-postage paid at Brooklyn, N.Y., and at additional mailing offices. Subscription included in Botanic Garden membership dues ($20.00) per year).

PLANTS & GARDENS

BROOKLYN BOTANIC GARDEN RECORD

SOILS

THIS HANDBOOK IS A REVISED EDITION OF PLANTS & GARDENS, VOL. 42, NO. 2

HANDBOOK #110

There are thousands of soil types, each with different abilities to supply plants with water and nutrients and allow roots to penetrate deeply for support and drought resistance.

Photo Courtesy USDA Soil Conservation Service

FUNCTIONS OF SOIL

ROY L. DONAHUE, JOHN C. SCHICKLUNA
AND LYNN S. ROBERTSON

From soil plant roots receive mechanical support, essential nutrient elements, water and oxygen.

Soil *physical* properties largely determine a soil's water-supplying capacity to plants. Soil *chemical* properties determine the soil's nutrient-supplying capacity. Both physical and chemical properties determine root extension and the volume of soil that serves as a reservoir for both water and essential nutrients for plants.

Some soils are in a physical and chemical condition that encourages plant roots to grow deeply and extend

long distances laterally. These soils are ideal because the plants growing in them will not blow over, will be drought resistant and capable of absorbing nutrients from a large volume of soil. The growth of plants can be restricted by naturally or artificially compacted or infertile layers, too much or too little soil moisture or soluble salts in toxic quantities.

At present sixteen elements are known to be essential for the growth of most plants: carbon, hydrogen and oxygen from air and water; phosphorus, potassium, sulfur, calcium, iron, magnesium, boron, manganese, copper, zinc, molybdenum and chlorine from soil; and nitrogen from both air and soil.

ESSENTIAL ELEMENTS FOR PLANTS FROM AIR AND WATER

Carbon (C), Hydrogen (H), Oxygen (O), Nitrogen (N)

ESSENTIAL ELEMENTS FOR PLANTS FROM SOIL AND FERTILIZER

MAJOR NUTRIENTS	Nitrogen (N) Phosphorus(P) Potassium (K)
SECONDARY NUTRIENTS	Calcium (Ca) Magnesium (Mg) Sulfur (S)
MICRO-NUTRIENTS	Iron (Fe) Boron (B) Manganese (Mn) Copper (Cu) Zinc (Zn) Molybdenum(Mo) Chlorine (Cl)

FROM ***Soils: An Introduction to Soils and Plant Growth*** by Donahue/Shickluna/Robertson; 1971. Reprinted by permission of Prentice-Hall, Englewood Cliffs, NJ.

Water and air occupy pore spaces in the soil. Following a heavy and prolonged rain or irrigation, soil pores may be almost completely filled with water for a few hours. After a day or two, some water will have moved downward in response to gravity, and the larger pores will be emptied of their water but filled with air. With a further loss of water by evaporation or transpiration, air will replace more of the space occupied by water. The next soaking rain, or irrigation, repeats this process.

The important part of this air-water relationship is that there must be enough total pore space and that the pore spaces must be the proper size ranges to hold enough air and water to satisfy plant roots between cycles of rainfall or irrigation.

From the soil plants accumulate nitrogen, phosphorus and sulfur. In other words, plants nearly always contain a higher percentage of these elements than the soil in which the plants are growing. Conversely, soils almost always contain more iron, calcium, potassium, magnesium and manganese than the plants growing in them.

Essential Elements for Plants

Nitrogen was first proved to be essential for plants in the eighteenth century; seven more elements were proved essential in the nineteenth century; and thus far in the twentieth century, eight additional elements have been shown essential. As a result of more refined research methods, other elements probably will be proved necessary.

Plants contain sodium, iodine, selenium and cobalt, which have not yet been proved essential for plants but are a necessity for the people and animals who eat these plants. Although sodium increases the growth of certain plants, it has not been demonstrated as essential. Silicon and aluminum also occur in all plants, but apparently they serve no useful function. Throughout the world today, soil is the principal supply source of essential elements for plant growth. Fertilizers and manures are used only as a supplement. Green-manure crops function primarily to make existing soil nutrients more readily available.

Soil is the source of thirteen of the sixteen elements essential for plant growth. Twelve of the thirteen elements originated in the parent rock from which the soil developed. (Nitrogen was present in some parent rock in very small quantities.)

The total supply of plant nutrients in the soil is therefore of fundamental importance to the economic production of plants. Although it is true that plants can be grown in pure quartz sand when essential water, air and nutrients are supplied, economic production is seldom possible by this technique.

FERTILIZING PERENNIALS

(These Guidelines Can Also Be Used For Annual Flowers)

After years of declining popularity, herbaceous plants are making a fast comeback. Recent years have brought a surge in gardening interest, and perennial plants are being used more often in borders and mixed plantings. As people spend more time gardening with perennials, they are developing a genuine interest in learning good cultural practices. Long-term success with perennials, just as with woody plants, depends upon providing a suitable root environment. If other cultural requirements are met, perennials will benefit from proper fertilizer applications. Guidelines for fertilizing perennials are as follows:

In early spring (February or March) use a complete slow-release fertilizer with a 1:1:1 ratio, such as "10-10-10." Apply a surface application of fertilizer at the rate of 1 lb. of nitrogen per 1,000 sq. ft.

By late May, weak solutions of liquid fertilizer applications may be applied frequently. A highly soluble fertilizer, applied in two-to three-week intervals from late May through late July is suggested.

Read the label and adjust the application rate according to how often the fertilizer is applied. (The more frequent the application of fertilizer, the weaker the solution should be.) If the perennials have lush foliage growth and few flowers, reduce or eliminate the liquid fertilizer applications.

Stop fertilizing in late July and apply one more application of slow release fertilizer in late fall. Apply at the rate of 1 lb. of nitrogen per 1,000 sq. ft. This late application will promote root growth before the ground freezes hard and will increase productivity the following season.

Mary L. Wilmoth
Propagator; The Dawes Arboretum

PHYSICAL PROPERTIES OF SOIL

JOHN PAUL BOWLES

Thousands of soil types exist. Soils are classified according to a complicated and detailed system not unlike that used to arrange and show relationships between plants. Phylum, Class, Subclass, Order, Family, Genus and Species are the terms (broad to specific) used to name plants. Order, Suborder, Great Group, Subgroup, Family, Series and Phase are used to name soils.

These classifying divisions tell soil scientists the origin of the soil — what type of rock the soil is derived from and how the soil particles got there (wind, water movement, etc.) — about the natural plant cover, how deep is the soil, what type of horizons or layers are present, physical properties (textural qualities, organic content, nutrient content, structure) and more. Most of the information pertaining to soil classifying divisions is useless to home gardeners because it is so filled with unfamiliar terms and definitions. But at least the last or most specific category — Phase — is easier to explain and more useful for gardening decisions.

JOHN PAUL BOWLES *is Horticulturist at The Dawes Arboretum, Newark, Ohio. He is Guest Editor of this handbook.*

Phase, also called Soil Type, is not officially recognized in the United States system of soil taxonomy, in part because it repeats analyses of physical soil properties included in other categories. These soil properties, however, are most measurable and useful to ornamental and food-producing gardeners. A description of a Soil Type or Phase includes the kind, size and percentage of soil particles (sand/silt/clay), percentage of organic matter, references to mineral ion content (high aluminum, low phosphorus, etc.), soil consistency (the amount of resistance to tilling) and soil structure (the kind and amount of soil particle aggregates).

Land is also classified by its agricultural capability. Most of the earth's surface is covered by water. Much of the remaining area cannot (or should not) be cultivated because it is too poor or minerally unbalanced, too dry, too rocky, too steep, and so on.

Only about two percent of the land in the U.S. is ranked as Class I in land capability. Class I land is relatively flat and well-drained with rich, deep, well-structured soils. But with good management and/or irrigation, between 20 to 42 percent of the remaining land is suitable for regular cultivation. The U.S. is unusually well-favored in this regard, for in many countries, particularly those in tropical climates, the percentage of even marginally arable land is very, very low.

What Is Soil?

Before discussions of soil management can begin, just what soil is must be determined. Soil is in the upper portion of a part of the earth's crust called the regolith. The regolith is a layer of unconsolidated debris (boulders, rocks, stones, gravel, sand and the even more weathered materials, clay and silt dust) found above bedrock.

In some places the regolith is very thick, in others very thin or even absent. But in general, the regolith covers land in a coat thick enough so that the soil is from three to six feet deep. *Soil* becomes differentiated from the regolith by the weathering actions of wind, water and heat, and because plant roots, surface organic deposits (plant and animal remains) and the activities of animals and soil microorganisms have added organic matter. Just as weathering and organic matter sort soil from the regolith, more weathering and a high quantity of organic matter make *topsoil.* All plant roots are found in soil and most are found in topsoil.

Most of the world's soils are mineral soils because the organic content, even in the topsoil, is very low — usually five percent or less. The rest of the bits and pieces that make up "dirt" are soil particles (various and sundry minerals).

When describing soil, these particles are called *soil separates*: sand, silt and clay. Sand particles are the largest, clay the smallest; in fact, individual clay particles are microscopic. The percent comparison of soil separates decides how a particular soil textural type, such as sandy clay, loam and loamy soil, is named.

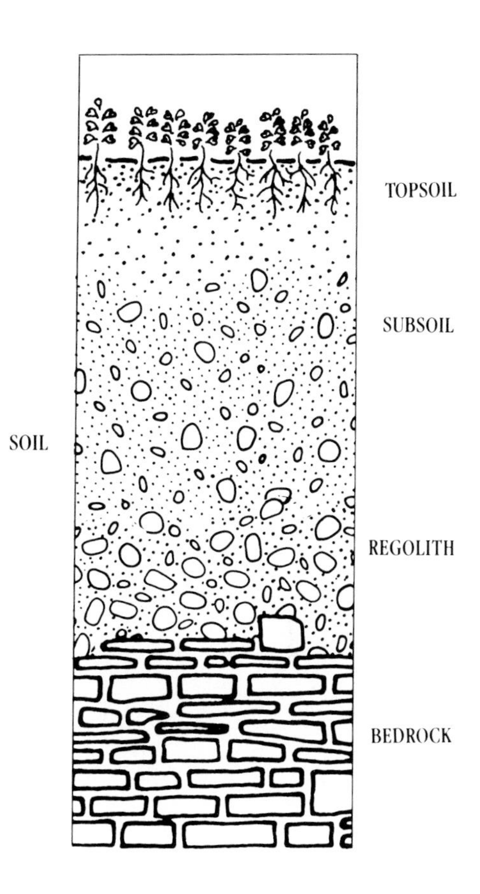

The regolith over much of the earth is very deep, but soil and topsoil vary considerably in depth. Sometimes the regolith is entirely soil, in which case it is usually thin and rests directly on bedrock.

Soil consistence refers to the resistance of a soil to rupture, i.e. tilling. Soil consistence is described in different ways depending on soil-moisture conditions. *Friable,* a term every gardener has heard, refers to a moist soil that crushes easily with gentle pressure but will hold together when pressed. This is the best condition a soil can be in, both for plant growth and cultivation. Soil *tilth* is often referred to in gardening and agricultural tests; tilth is the physical condition of a soil as related to its ease of tillage, fitness as a seedbed and impedance to seedling emergence and root penetration. A soil with good tilth and friability makes for a very good garden.

Consistency, tilth and friability are dependent on soil texture and on *soil structure.* The structure of a soil refers to the way in which its individual particles cling or *aggregate.* Soil structure is a very important quality because it determines how effectively air and water move into and through the ground.

If a clay soil is non-aggregated, spaces between particles, called *pores,* are so tiny that water, and even air, will not move or *percolate* (flow) down, and plant roots "suffocate" from lack of air or too much water. If a very sandy soil contains no aggregates, then there is nothing to impede the flow of water, and plants suffer from drought. Naturally formed aggregates are called *peds*; unnaturally formed aggregates, such as are made when tilling a too-wet clay soil, are called *clods.* Clods are undesirable mainly because their formation ruins natural soil structure by mashing peds into too-large clumps. Aggregation that's best for plant growth is present in a soil that appears grainy, not chunky; desirable soil aggregation occurs almost exclusively on a microscopic level.

Garden plants require both water *and air* at their roots. Clay soils, because of a multitude of tiny particles, have more total pore space – pore space in the "blank" area between soil separates – than sand soils. But the pores in clay soils are often *too* small for good air or water movement. Water is held in great quantities in clay soil pores but much of it is unavailable to plants. Sand soils, on the other hand, may have excellent drainage and good aeration, but are not able to store water. So in a way, very clayey and very sandy soils have the same problem: poor pore structure; sandy soil is too loose and dry, and clayey soil is too tight and airless. For home gardeners who must work with the soil they have, yet want to grow a variety of plants in favorable conditions, the management and improvement of soil is of the utmost interest and importance.

It is essential that you not overtill soil when preparing it for planting. A pulverized or powdery soil is *not* best for plant growth. And if you use a mechanical tiller, be advised that, as it breaks up topsoil, the pressure of its weight forms a compaction below the area being tilled – sort of a big, flat clod (or pan) that will resist water movement and plant roots.

Why is all this noteworthy to you? 1) For best growth of most ornamental and edible plants, a friable, well-drained, airy, yet moisture-rich, soil is desirable; loose soil with lots of "good" pore space, yet also lots of water holding capacity. 2) Most agricultural soils, and of course the soil around your home, are mineral soils that may be heavy, compacted or poor. 3) Any soil texture type that has a good structure is easier to till and promotes better plant growth. 4) You *can* do things to increase desirable soil aggregation. What? One thing that is easy to do is to disturb soil as little as possible. Native soils often have a decent structure, and a disturbed soil, if let alone, may eventually re-develop its original nature. Disturbances include more than tilling; normal foot traffic or as simple but constant an activity as mowing can cause enough soil compaction to qualify as disturbances. One positive step you can take to encourage appropriate aggregation is to add organic matter.

But how you do all this – evaluate your soil and decide how to manage and/or improve it – is another matter, and that's what this handbook is all about.

REPRINTED, WITH PERMISSION, FROM *THE DAWES ARBORETUM NEWSLETTER.*

DETERMINING SOIL TEXTURAL CLASS

How To Use This Diagram

Along the labeled triangle sides, locate points corresponding to percentages of silt, sand or clay present in a soil under consideration. From each of those points, draw a straight line inward, parallel to the side adjacent in a counterclockwise direction. The name of the compartment in which the lines intersect becomes the class name of the soil in question.

Using the "Feel" Method to Determine Textural Class

When a small amount of dry soil is rubbed between forefinger and thumb:

SAND feels gritty and is not moldable or sticky.

Dry **SILT** has a floury or talcum-powder feel, is only somewhat moldable and sticky when wet; dry clods are somewhat difficult to crush.

CLAY is very moldable when wet; dry clods are extremely difficult to crush.

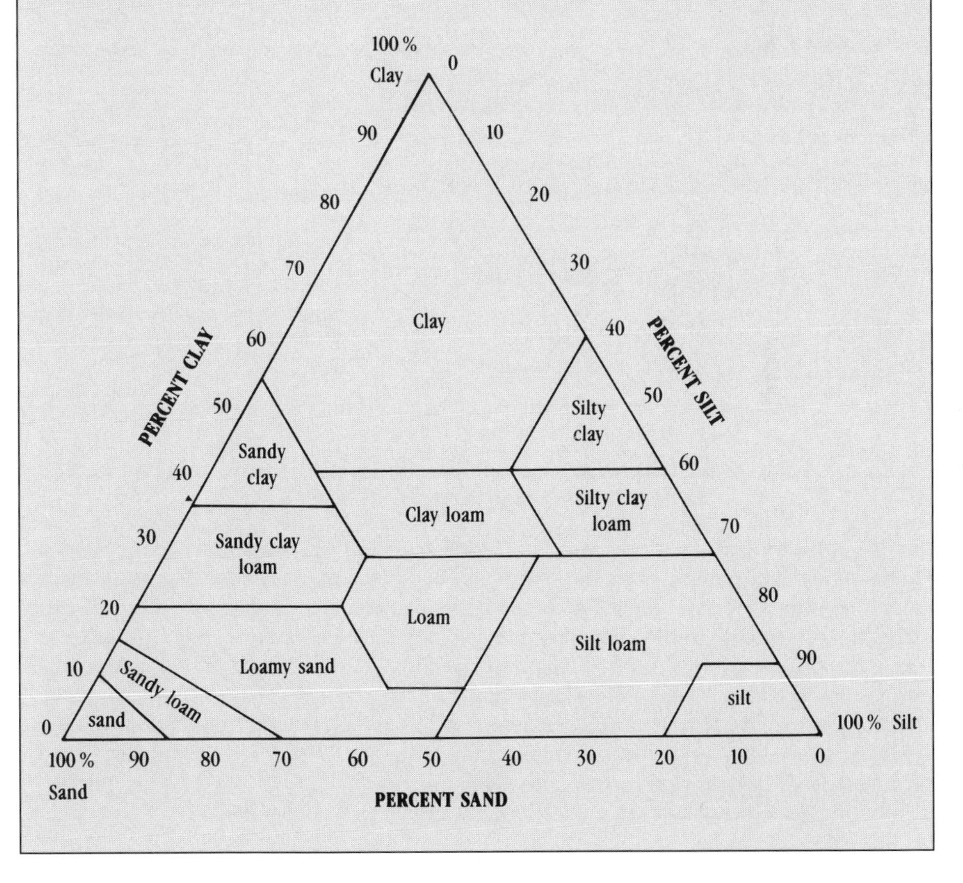

General Terms Used to Describe Soil Texture in Relation to the Basic Soil Texture Class Names

Sandy Soils	Loamy Soils			Clayey Soils
Coarse	Moderately coarse	Medium	Moderately fine	Fine
Sandy Loamy sands	Sandy loam Fine sandy loam	Very fine Loam Silt loam Silt	Clay loam Sandy clay loam Silty clay loam	Sandy clay Silty Clay Clay

Graphic Guide to Textural Classification

Because of the small scale of the chart, it is not possible to show all recognized soil textures in the sandy range.

Soil Separate	Diameter Range (mm)	Comparison to U.S. Monetary System (Dollars)
Very Coarse Sand	2.0 -1.0	2.00-1.00
Coarse Sand	1.0 -0.5	1.00-0.50
Medium Sand	0.5 -0.25	0.50-0.25
Fine Sand	0.25-0.10	0.25-0.10
Very Fine Sand	0.10-0.05	0.10-0.05
Silt	0.05-0.0002	0.05-
Clay	less than 0.002	

Mineral fragments in the soil larger than two millimeters in diameter (gravel) are not strictly a part of the soil, but must be recognized because they greatly influence the use of land. For example, a sandy loam that contains a larger amount of gravel becomes a gravelly sandy loam.

Organic soils are rather rare. Soils are classified as *organic* when organic content is between 20 and 30 percent by dry weight. Considering the light weight of dry organic matter, a true organic soil contains a very, very large amount of organic material. (A soil test can tell you the percent of organic matter in a soil sample.)

Lawn Clippings

To Remove or Not To Remove

Must you remove lawn clippings? *Only* when the clippings are thick enough to mat together and smother the grass beneath them. But then, you should never remove more than one inch of grass when you mow, and one inch of clippings will not smother grass. In fact, when grass is mowed properly, the next day after mowing, clippings will seem to have disappeared. Why? Because clippings are over 90-percent water and quickly dry up to almost nothing. It's for this reason that clippings do *not* contribute significantly to thatch. Many other cultural practices cause thatch, but the removal/non-removal of clippings has very little to do with it. So mow grass to the proper height—two inches or more for most lawn grasses—and mow often enough so that clippings won't be a problem. Note: Uncollected clippings return nutrients and organic material to the soil.

SOIL ORGANISMS

STEVEN J. THIEN AND
BARBARA DANIELS HETRICK

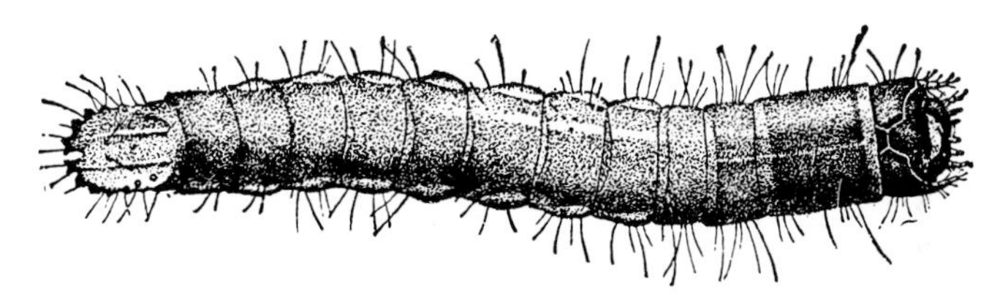

Wireworm

The soil environment supports a vast amount and array of organisms. Their importance depends on each organism's ability to influence soil properties and plant growth. Without the collective activities of the multitude of organisms it harbors, soil would be an inert medium incapable of supporting unassisted plant growth. Instead, a productive soil constitutes an ecosystem alive with organisms, seen and unseen, responding to the energy level and climatic limitations of their unique environment.

Types of Soil Organisms

In addition to miles of plant roots, almost innumerable organisms populate soil. These include both plants (flora) and animals (fauna), which can be further categorized into micro- (small) or macro- (large) sized types. Although macroorganisms are most noticeable, the microorganisms exert a more pronounced influence on vital soil processes. Microorganisms are responsible for organic matter decomposition, nutrient cycling and other wide-ranging effects on plant growth.

MACROFAUNA. Nematodes are by far the most numerous of soil animals, but even though classed as macro-sized, are still too small to be seen. Other macrofauna include: snails, slugs, earthworms, woodlice, springtails, millipedes, centipedes, termites, beetles, ants, spiders and mites. Macrofauna are involved in the initial stages of the breakdown of surface organic residue and its incorporation into the soil. Other animals, such as rodents and snakes, inhabit the soil

STEVEN J. THIEN *is Professor of Soil Science at Kansas State University, Manhattan, Kansas.*

BARBARA DANIELS HETRICK *is Assistant Professor in the Department of Plant Pathology, Kansas State University, Manhattan, Kansas.*

for a part of their existence, but are less important to soil dynamics.

MICROFAUNA. Soil harbors few of these organisms. Protozoa are the most notable; mainly they parasitize other soil organisms.

MACROFLORA. Plant roots are the most obvious in this group. The metabolic activity of roots has a marked effect on soil properties and, hence, on the activities of other organisms.

MICROFLORA. This category of soil organisms is the most influential. Members include: bacteria, fungi, actinomycetes, algae and viruses.

Bacteria are single-celled organisms occupying favorable niches as colonies of many cells. In a rich environment, 1,000,000,000 bacteria cells may occupy a cubic centimeter of soil.

Fungi usually account for the largest portion of the total soil microfloral mass. Fungi permeate the soil as threadlike networks of filamentous material called mycelia. One gram of soil can contain as much as fifty meters of fungal mycelia.

Actinomycetes produce single-celled colonies like bacteria, but also permeate the soil with slender mycelia. Actinomycetes normally compose ten to fifty percent of the microfloral soil population, but may dominate the microflora population in dry, alkaline conditions. The familiar aroma of freshly-turned earth originates from these organisms.

Algae possess chlorophyll and need access to sunlight, so populations are confined to the surface layer of soil. Their inability to compete with other organisms except under certain circumstances limits their influence to very wet conditions. Blue-green algae, however, are some of the few free-living organisms that can incorporate atmospheric nitrogen into plant-available organic compounds.

Viruses are the smallest soil organisms, being perhaps twenty times smaller than bacterial cells. They rarely exist free in the soil, but instead invade other soil organisms, frequently becoming pathogenic (disease producing).

Soil Organism Activity

The amount of carbon in soil determines the energy available for soil organism activity. Nowhere is this relationship more evident than in the process of organic matter decomposition.

Plant residue, when added to soil, is changed into humus by the chemical actions of soil organisms. First, macroorganisms reduce the size of the residue material and mix surface residues into the soil. Then microflora, predominantly bacteria and fungi, further decompose the residue by extracting energy stored in molecular chemical bonds. This process releases nutrients and carbon dioxide as by-products. The original residue is by now converted to a new compound called humus.

Humus is nitrogen-enriched when compared to the original residue, and gives soil a dark coloration. Humus acts as a storehouse for both water and nutrients; humus granulates and loosens soil, encouraging better root, air and water penetration. And humus provides a carbon energy supply that encourages beneficial microfloral activities.

During organic matter decomposition, soil organisms release nutrients from residues and make them available to plants. Most notably, nitrogen, phosphorus, sulfur and boron are converted from complex organic forms to the simple mineral forms required for plant growth. Through this series of processes, called mineralization, the recycling activities of soil organisms increase soil fertility.

The small zone of soil immediately adjacent to a plant root abounds with soil organisms. This zone contains microfloral activity ten times (or even several hundred times) that observed in soil farther away from roots. The intimate association between plant roots and soil organisms provides the ideal environment for plant and soil organism interaction. For example, many soil organisms produce growth-regulating compounds that stimulate plant growth; other substances act as toxins and are harmful to growth. Some microorganisms act as a sort of defensive shield and suppress damaging organisms. All soil microorganisms utilize enzyme systems to aid specific chemical reactions, which are often beneficial to plant roots. Once understood, the regulation of plant root and microorganism interaction holds promise as a tool in soil management for plant production.

Soil Management and Soil Organisms

In general, a healthy soil is characterized by an abundance of soil organism activity. Gardeners should strive to keep soils well supplied with carbon-containing residues (dead plant material) to maintain active beneficial soil microorganism populations. Rotating plant types on a particular site is also beneficial, as this practice prevents the build-up of undesirable organisms.

The key to suppressing disease organisms is to encourage and maintain good plant health. Plants experiencing moisture, temperature, fertility or aeration (compaction) stress are more vulnerable to disease or pest invasion. Growers should develop and maintain soils that provide ample drainage and aeration. Too low/high pH or nutrient deficiencies/toxicities should be corrected and organic residues incorporated to stimulate soil organism activity.

Controlling Undesirable Organisms

At some time most growers are faced with controlling undesirable soil organisms. The approach and chance for success varies with the type of organism.

Macrofauna Control. Small-mammal control is difficult. Poisons, fumigants and traps are available, but a successful program may have to include more than one technique. Burrowing animals, such as ground squirrels and moles, can cause considerable damage by building extensive tunnel systems. These tunnels also provide soil access to field mice and insects.

Poisoned food baits are effective for squirrels and mice, but are extremely toxic and should be used with caution. Moles feed mainly on earthworms and grubs and are not easily controlled with bait poisons. Controlling grubs in lawns forces moles to look elsewhere for food; or they can be exterminated with traps or poison-gas cartridges discharged in their tunnels.

Nematode Control. Nematodes are microscopic, wormlike animals that inhabit nearly all soils. Most nematodes feed on bacteria, but a few species possess a hollow, needlelike mouthpart called a stylet, which is used to pierce roots and extract plant juices. While feeding, plant-parasitic nematodes inject a chemical which causes roots to swell or knot and develop galls or lesions.

Excessive branching of roots or stubbiness are also signs of nematode feeding. The wounds caused by feeding nematodes provide entry sites for bacteria and fungi, and in this manner nematodes facilitate root rot and other diseases.

Above-ground symptoms of nematode damage are reduced growth, yellowing of leaves excessive wilting and reduced yields. But since symptoms of nematode damage can be similar to symptoms of nutrient deficiencies or disease, it is always advisable to seek professional diagnostic advice.

The objective of nematode control is

to reduce the population of nematodes to a point where adequate plant growth and yield is achieved. Cultivation helps reduce nematode populations by injuring them and exposing them to drying by the sun and wind. Nematode control is aided by removing the host plants through fallowing, or planting a nematode-resistant plant for one season or more. Early crop-residue disposal, the use of resistant varieties and crop rotation may starve the pests. Asparagus and marigolds release toxic substances that decrease some nematode populations. Steaming or heating small quantities of potting soil at 350 degrees Fahrenheit for one hour effectively kills nematode populations; however, nematodes are rarely a problem in houseplants.

In addition to cultural practices, nematicides (chemicals toxic to nematodes) offer a control option. Nematicides must be applied to bare soil, but can greatly reduce nematode populations. Some nematicides also control fungi, bacteria, insects or weeds. Unfortunately, the most effective agents, such as methyl bromide, are highly dangerous and should be used only with the recommendation and supervision of specialists.

Soil-borne Pathogen Control. Some soil organisms (certain types of fungi, bacteria and viruses) can invade roots and become pathogens (disease-causing organisms). They pose special problems to growers since they are almost impossible to detect and rank as the most difficult of all disease agents to control. Soil-borne diseases are responsible for most crown and roots rots, wilts, blights and smuts. These organisms are very common and most gardeners are at some time faced with their control. The most serious diseases usually spread rapidly and appear year after year. For these diseases, control is aimed at prevention rather than cure and usually involves a combination of practices.

The best control for a disease is to restrict it from entering new regions. Once introduced, pathogens can quickly reach epidemic proportions. Federal and state laws regulate the importation and distribution of soil and plant materials to eliminate just this possibility. Home growers trading plants should likewise beware of introducing disease organisms.

In the home garden, good cultural methods for plants afford some degree of control by creating unfavorable conditions for plant pathogens. And plants in good growing environments are more resistant to pathogen damage. The spread of disease organisms is restricted where soil management encourages rapid organic-material decomposition. Practicing proper sanitation by promptly removing and destroying infected plant material and weeds may interrupt the life cycle and decrease the population of a pathogen.

Use of disease-resistant varieties offers the single most efficient method of biological disease control. Some pathogens cause disease only in very specific groups of plants. By growing plants that are unsuitable hosts, pathogen populations can be decreased or in some cases totally eliminated. A new biological control strategy being investigated uses beneficial microorganisms to regulate pathogenic organisms. Some day this may be a common method of combating disease.

Heat is an effective method for controlling undesirable soil organisms. Steaming (using moist heat) soil to an internal temperature of 190 degrees Fahrenheit for at least thirty minutes kills most insects, weed seeds and soil pathogens. With dry heat, one hour at 350 degrees Fahrenheit is required. But both these methods are not feasible for home gardeners and can be damaging to the soil itself. However, in some climates, solar heat can be used to "steam" soil if the soil is kept moist and covered with transparent plastic during the hottest period of the year. If the soil can be kept around 120 degrees Fahrenheit for two weeks, pathogens that cause wilts and root rots may be controlled.

It is important to note that steaming not only destroys soil organisms (good and bad), but also alters some physical and chemical properties of the soil. Heavy clay soils may become more granular with steam, often improving drainage and aeration. After steaming several applications of water may be necessary to moisten the soil thoroughly. In highly organic soils, steaming can cause ammonium nitrate to accumulate in quantities large enough to burn roots and foliage of sensitive plants. Other minerals may also increase toxic levels if soils are over-heated. Avoid repetitive heat treatments, excessively high temperatures and longer-than-necessary periods of heat.

Chemicals provide another means of disinfecting soil seeds and bulbs and controlling plant diseases or the insects that carry and spread diseases. As a rule, chemicals applied to plants seldom cure diseases brought on by soil organisms; rather, they are protectants that reduce or prevent the *spread* of disease. Seeds, seedlings and bulbs can be treated with chemicals to protect them from soil pathogens. Examples of such chemicals are zineb, maneb, thiram and chloroneb. Damping-off, seedling blight and crown and root rots can be controlled with soil-applied fungicides, such as those just listed.

Soil can be treated directly with another class of chemicals, called fumigants, that control some pathogenic fungi, bacteria and nematodes. Some fumigants disperse slowly through the soil, others are highly volatile and the soil must be covered with plastic to prolong their effect. Common fumigants are chloropicrin, methyl bromide, and vapam; all are very dangerous and should be used with extreme caution.

Fumigants are very potent chemicals, toxic to plants, animals and people. They kill all forms of vegetation, so soil near desirable trees, shrubs or other plants cannot be treated. Fumigants vary in their ability to move through the soil, but all require loose and moist conditions for optimum effectiveness. Most have waiting period of one to four weeks before safe seeding or replanting can take place.

Most gardeners find diseases difficult to diagnose. But the specific pathogen must be identified before an effective control program can be implemented. In most cases, recommendations on chemical and cultural control practices should be sought from specialists.

White grub

MYCORRHIZAE

They Can't Help but Help Plants

Gary W. Watson

Plant root systems cannot penetrate every area of the soil. Even the smallest root tips are much larger than soil micropores. Soil micropores are the tiniest of soil spaces and surround individual soil particles. These microscopic openings make up the bulk of soil pores and could provide access to tremendous amounts of nutrient ions if plant roots were small enough.

A symbiotic (or mutually supportive) relationship that overcomes this problem has evolved between plant roots and soil fungi. Certain fungi and plant root systems grow together, each becoming an extension of the other. The fungi produce structures that are called mycorrhizae. Mycorrhizae are not roots, but for ease of comprehension can be thought of as such. They go unnoticed because they usually cannot be detected with the naked eye. On a few plants, the bright color of the fungus can be observed as it covers the surfaces of plant roots.

Several types of mycorrhizae exist, each differing in structure, but with similar function. In each type the fungus actually grows from the soil into the fine root tips of a plant and establishes a mechanism for nutrient exchange between the two organisms. The fungus uses the plant as an energy source, receiving sugars that have been manufactured by photosynthesis in the leaves and transported down to the roots. The plant receives nutrient ions in return.

When mycorrhizal relationships are present, a network of extremely fine fungus strands extends from the root for many inches into the surrounding soil. This fungal network acts as an extension of the root system, penetrating areas of the soil that the roots cannot reach and absorbing and transporting nutrients back to the plants.

In nature only a few plants species and special soil situations are known where plant roots totally lack mycorrhizae. But site factors do influence the degree of mycorrhizal association. On infertile soils, growth of mycorrhizal plants far exceeds nonmycorrhizal plants. But on sites of high fertility, the effectiveness of mycorrhizae diminishes, and may be poorly developed. Also, wet sites have less mycorrhizae than dry sites.

At this time there are no reliable procedures or aids known to encourage mycorrhizae and their benefits. But research on this topic is increasing and someday the management of mycorrhizae may be as important as the use of fertilizer.

Garry W. Watson *is Plant Physiologist at the Morton Arboretum, Lisle, Illinois.*

LIVING THINGS IN THE SOIL

What They Do for You, What You Do for (and to) Them

John Paul Bowles

As discussed in the previous article, *Soil Organisms*, the ground under our feet is full of life. So many things live there you almost expect the earth to churn and squirm.

Many organisms live in soil: bacteria and various other kinds of microorganisms (living things too tiny to see with the unaided eye), higher plants (roots mainly), earthworms, ants and other arthropods, beetles and their larvae (grubs), other kinds of insects, slugs, snails, nematodes, fungi (mushrooms, puffballs, mycorrhizae, etc., and many microscopic forms such as molds and yeasts), protozoa (microscopic animal life), mice, moles, groundhogs and who knows what else? A handful of soil may contain billions of organisms.

Both animal and plant life depend on soil organism activity.

Bacteria are the most numerous and arguably most important soil organisms. There are many kinds of bacteria. Some of the most beneficial types convert atmospheric nitrogen (all plants require nitrogen to live) into forms plants need. This is called nitrogen fixation — something plants cannot do for themselves. Some plants (such as peas and other legumes) have nitrogen-fixing bacteria that live right in their roots.

Other bacteria, and another group of microorganisms called actinomycetes, are responsible for the decomposition of dead things, a useful task, if not for the sake of good housekeeping, then certainly for the advantages of recycled nutrients.

Some soil microorganisms have antibiotic qualities. Penicillin and Streptomycin, familiar because of their use as human medicine, are produced in quantity by soil fungi and actinomycetes. It is thought that soil antibiotics help control some plant diseases.

The activity of decomposition mi-croorganisms provides very useful byproducts: carbonic, nitric and sulfuric acids. These acids are essential in both the formation of soil and for plant growth, as they dissolve rocks and turn minerals needed for plant growth from insoluble (unusable) to soluble (usable) forms. Besides slowly dissolving rocks into smaller particles, many types of soil microorganisms and their by-products aid in the "cementing" of tiny soil particles into bigger aggregates, and thus improve soil tilth.

Many other creatures such as ants, centipedes, various mites, springtails, earthworms and nematodes (a kind of microscopic animal) participate in the breakdown of dead material. And the physical movement of most of these animals, particularly earthworms, is

A good mulch will encourage desirable soil organisms.

PHOTO BY ELVIN MCDONALD

important for soil building and conditioning. Mammals that live in-ground also contribute to the development of better soils for plant growth. Plants themselves help build better soils as their roots grow, loosening heavy matter and even breaking stones. Plants also contribute a large quantity of the dead material required for the life of most beneficial soil organisms.

So healthy soils are teeming with life, but all things that live in soil are not conducive to the health and welfare of plants. There are soil-borne disease organisms, and most people are aware of the detrimental feeding effects of some kinds of nematodes and the grubs of beetles. Anyone with a lawn has been irritated by the tunnels of moles (which, by the way, almost exclusively eat the grubs that almost exculsively eat grass roots).

Is there any way to encourage "good" soil organisms while discouraging "bad?" Not so very long ago the use of non-selective soil fumigants and sterilants was often advised for garden soils, and sometimes still is for particularly severe problems. However, these products usually do more harm than good by also killing the multitude of desirable organisms. Selective chemical treatment for undesirable pest or disease control is better than a soil sterilant. For example, use a selective nematicide to get rid of nematodes (but only when you are *certain* that you have a severe nematode problem).

An even better method to encourage desirable organisms at the expense of undesirables is thoughtful and timely management. Remove or treat diseased plants promptly, and grow plants tolerant of the site (a plant that prefers sun but is planted in shade is prone to more disease problems). Keep soil pH in the good growth range (5.5 to 6.8), and nutrients at desirable levels (use soil tests!). Mulch whenever possible and manage lawns so that clippings don't have to be removed and dead organic material (that becomes microscopic food!) is present.

The management of soil pH, nutrients (particularly nitrogen) and organic matter is vitally important to the well-being of the kind of soil organisms (including earthworms) that build or maintain healthy soils. A healthy soil is good for plants and good for you.

REPRINTED, WITH PERMISSION, FROM *THE DAWES ARBORETUM NEWSLETTER*.

LIVING QUARTERS FOR PLANT ROOTS

HENRY C. DE ROO

Soil is the living quarters of an important part of most plants, the roots. Frequently half of a plant or even more, consisting entirely or largely of roots, lives and grows in the soil.

HENRY C. DE ROO, *now retired, was soil scientist for The Connecticut Agricultural Experiment Station, Valley Laboratory, Windsor, Connecticut.*

The main function of roots is to anchor the plant and to supply it with the nutrients needed and enormous amounts of water, all of which come from the soil.

For most species of plants, there is a rather definite relation between root and top growth. It follows then that a well developed, healthy root system is essen-

Often half of a plant or even more, consisting entirely or largely of roots, lives and grows in the soil.

tial for the production of a vigorous plant.

These facts are so commonplace that there is danger of forgetting them. Even farmers and gardeners, though wise in ways of plants, have a habit of neglecting the roots. This is quite understandable, however, because most crops are commonly evaluated by the above-ground parts. Beets, carrots and turnips are among the few crops grown primarily for their roots.

In spite of the general lack of appreciation for the importance of plant roots, most efforts to promote top growth are actually directed to the roots of the plants. Fertilization, irrigation and most cultural practices affect the roots first. Exceptions, of course, are plant breeding and the control of plant pests and weeds.

It is remarkable how far or deep plant roots can go in search of nutrients and water. The root system of a single corn plant, for example, may occupy 200 cubic feet of soil. On young corn plants, in the five-leaf stage, 23 roots were counted which, in total, were crowded with 10,000 side roots. If all the roots of a single wild oat plant grown under favorable conditions and excavated 80 days after sprouting were put end to end, they

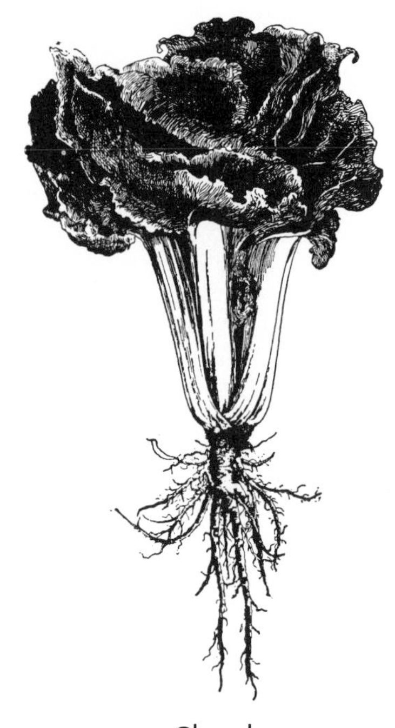

Chard

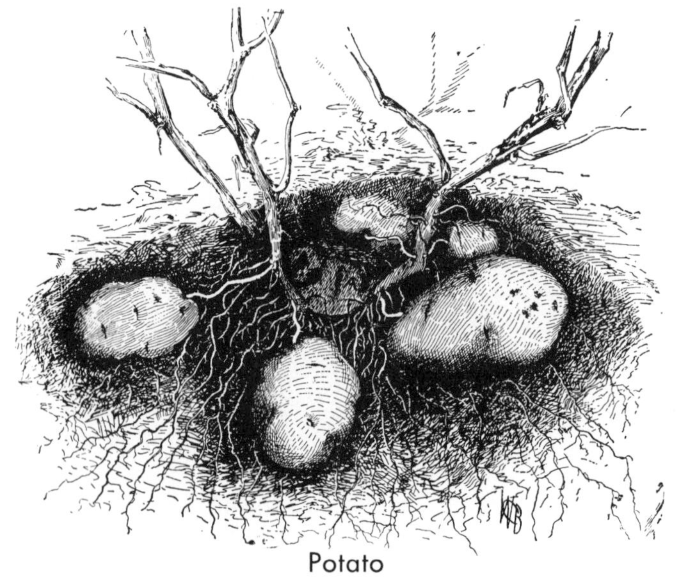

Potato

In the sketch at left, the soil was not prepared deeply enough, and the manure, left near the surface, caused crooked and branched root growth. Deep soil and thorough mixing of soil and manure are essentials in growing carrots and other root crops.

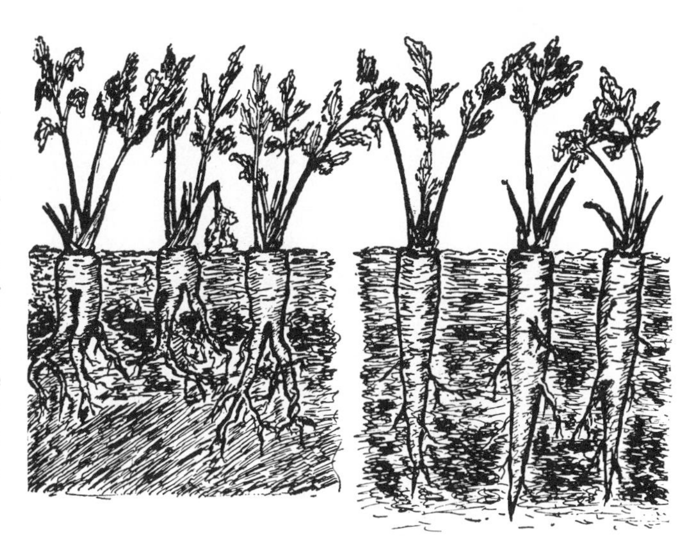

would measure approximately 54 miles.

An ideal soil provides a well distributed system of water reservoirs throughout the soil mass, enough air to permit adequate root respiration and microbiological activity and lots of spaces in the soil into which roots can grow and develop freely.

A "spade test" is one way of getting a look at soil. Dig a narrow pit with a spade and take a thin slice from one of the undisturbed walls. Gently shake this sample and shatter the soil. If it is moist and of good structure, it will come apart in rounded porous "crumbs." A poorly-structured soil or soil layer breaks apart into blocklike clods having flattened surfaces with vertices more or less sharply angular. Cloddy, lumpy soil with extremely fine, almost invisible pores

Turnip

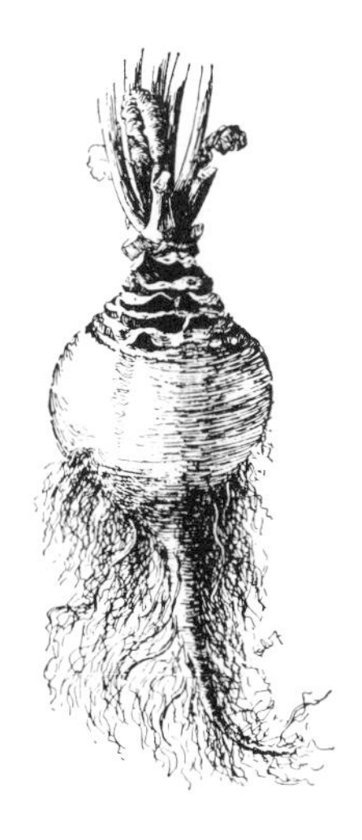

Rutabaga

Leek

within the lump is a pretty sure sign of soil structure deterioration and/or soil compaction. Such a soil is hard to work into a favorable seedbed and becomes waterlogged after every heavy rain.

Soil compaction is caused by pressure on the soil surface, especially when the soil is too wet to work, or to walk or drive on. Careless timing of tillage practices can easily do severe damage to a soil in good physical condition. To cure a bad structure or compaction pan takes a lot of time and effort. In severe cases, the surface soil or the layer of soil at plow or tillage depth may become almost as hard as a concrete floor. Use of a bulldozer can bring about such a condition.

What has been indicated here on the subject of soil-root relationships emphasizes the importance of examining your garden soil. Dig a little deeper than usual, perhaps two feet down, where the soil has a brighter color, and look for differences in color, compaction or soil depth. Judge for yourself whether or not your soil provides the best conditions for the growth of your plants. If you are not sure, consult an expert. Where soil improvement is too difficult and expensive, the alternative is to choose only those plants that can tolerate the soil conditions you have.

SOIL pH

CLYDE E. EVANS

Soil pH is used to indicate the relative acidity (or alkalinity) of a soil. pH is defined as the logarithm of the reciprocal of hydrogen ion activity. What this means is that the amount and type of hydrogen ions in soil water determine how acid the soil is. Ions are charged molecules, either positive or negative. Their charge and quantity affect the rate and outcome of soil chemical reactions. pH compares hydrogen ions (H+) to hydroxyl ions (OH-) in soil water, or the soil solution. When hydrogen or hydroxyl ions are in equal amounts, the pH is said to be neutral. When hydrogens prevail, pH is acid; when hydroxyls are most represented, pH is alkaline (sometimes called basic).

pH runs from zero to fourteen (acid to alkaline), with seven indicating neutral. pH is a logarithmic measure; that is, each increment is ten times as great as the next. pH 4.0 is ten times as acid as pH 5.0 and one hundred times as acid as pH 6.0, for example.

As pH becomes acid or alkaline (and free hydrogen and hydroxyl ions increase or decrease), chemical reactions occur involving hydrogen ions that either restrict or release *other* ions (including plant nutrients) into the soil solution.

CLYDE E. EVANS *is Professor of Soils in the Agronomy & Soils Department of Auburn University, Auburn, Alabama. He is also the Director of Auburn University's Soil Testing Laboratory.*

Factors Affecting pH

Generally, soils formed under high rainfall conditions are acidic and those formed under arid conditions are alkaline (or basic). Soils receiving approximately thirty inches or more of annual rainfall are likely to be acidic; soils with less than that are likely to be basic. Because of high annual rainfall, soils of the eastern United States are usually acidic and those of the low-rainfall areas of the western United States are basic.

Fertilizers may be highly influential in soil acidity, depending on the source of their nitrogen. Those containing ammonium (NH_4^+) as a nitrogen source are acidifying. Acidifying substances include ammonium sulfate, ammonium nitrate, urea and ammonium phosphate, and any of these may be found in mixed, complete fertilizers. A complete fertilizer contains all of the three most important plant nutrients: nitrogen, phosphorus and potassium.

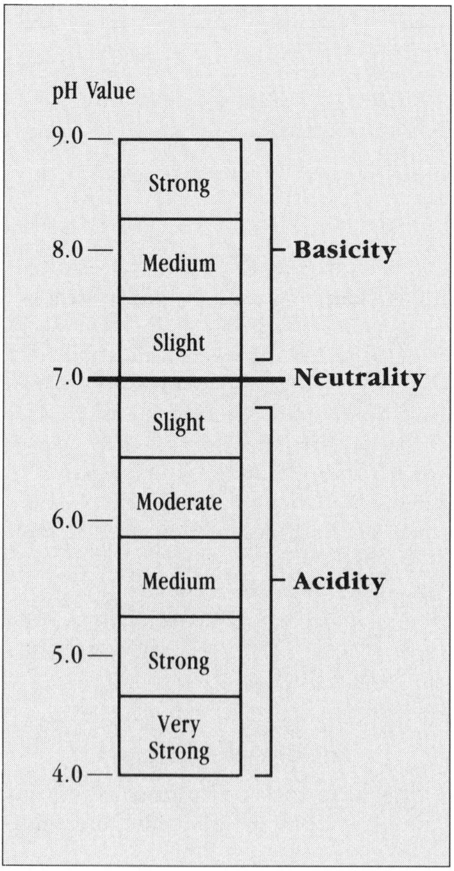

Sometimes soils are referred to as "sweet" or "sour." Generally, the term sweet is applied to a nearly neutral or alkaline soil and sour indicates an acid soil. Based on appearance alone, one cannot distinguish the acidity or alkalinity of a soil. Sometimes a soil containing organic matter may have a characteristic odor associated with the breakdown of organic material. Soils having such odors are often termed "sour" or acidic. While it is true that the decomposition of soil organic matter adds to acidity, odors associated with decomposition are not a direct indication of the degree of acidity.

pH Value
9.0
Strong
8.0
Medium
Slight
7.0
Slight
Moderate
6.0
Medium
5.0
Strong
Very Strong
4.0
Basicity
Neutrality
Acidity

Measuring pH

Two commonly accepted methods of measuring soil pH are indicator dyes and the pH meter. Indicator dyes are the least accurate but are useful in the field for making a rapid pH determination. With experience, this method can be fairly reliable. The more accurate and most-used evaluator, however, is the pH meter. With this device, a suspension of soil and distilled or deionized water is placed in contact with a glass electrode. A reading is then taken from a calibrated scale or digital readout. This is the method used by professional laboratories. Hand-held or portable pH meters are generally too inaccurate to be of any practical use.

Research has determined that most plants grow best in a pH between 6 and 6.8 (slightly acid). Additions of lime and sulfur contribute to chemical reactions that either tie up or release hydrogen or hydroxyl ions, and so affect pH. The change of pH to a level that provides the best mix and quantity of available ions for plant growth (generally between 6

pH Effect on Nutrient Availability and Microorganism Activity

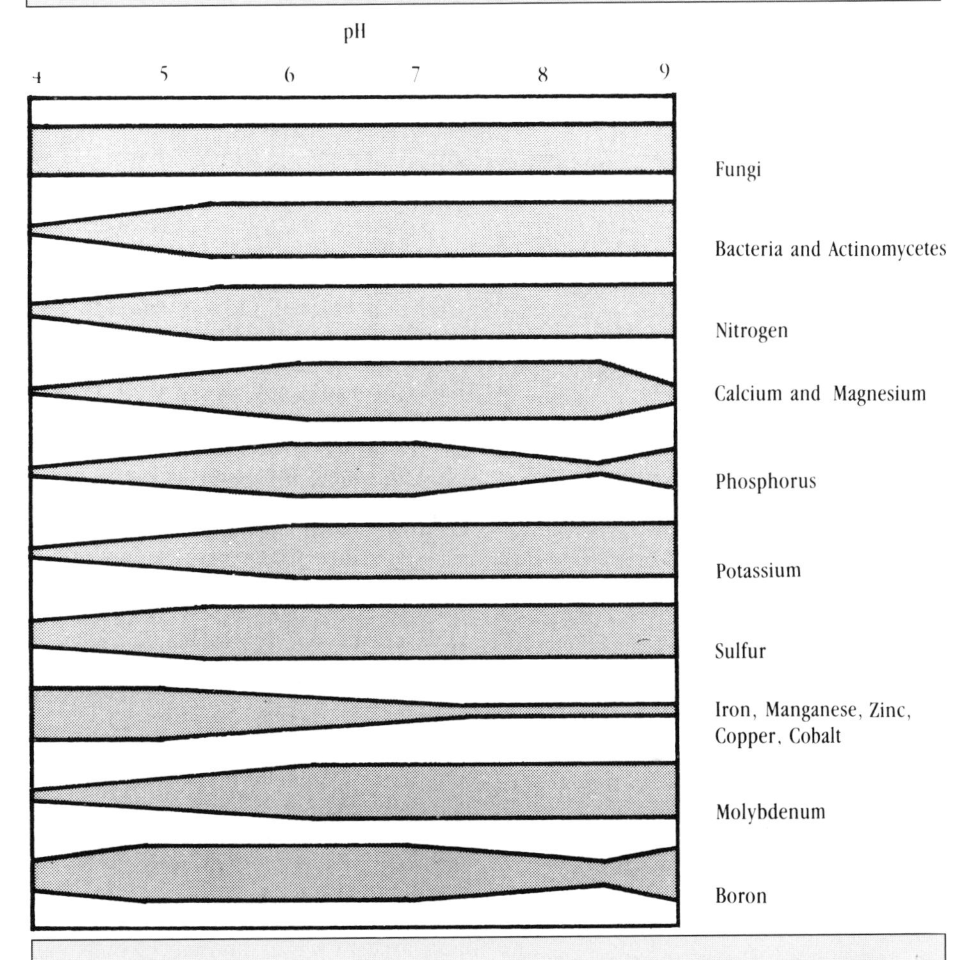

Shaded areas indicate nutrient availability. It is apparent that the majority are most available between a pH of 5.5 and 6.5.

and 6.8) is what is sought by the application of lime or sulfur. Since you must know existing pH before you can accurately change it (or determine if change is necessary), soil testing is important. Because soil make-up varies from area to area, even particle to particle, it is essential that you take many small soil samples and mix them together to obtain an average reading.

THE IMPORTANCE OF SOIL TESTING AND HOW TO DO IT

CLYDE E. EVANS

A lot of time, work and money goes into the establishment and maintenance of fine lawns and gardens. These efforts may be wasted if you lime and/or fertilize the soil improperly. Some plants require different soil acidity conditions than others. For example, azaleas, rhododendrons and gardenias can suffer from iron deficiency if pH is too high. Improper fertilization can cause plant-damaging phosphorous build-up, particularly in sandy soils. It can also cause deficiency problems. So just any fertilizer or fertilization program won't do.

Don't guess at soil lime and fertilizer needs. Many factors such as weather and insects affect plant growth. You can't predict these, but when it comes to a soil's nutrient status, you can take steps to determine current conditions and future fertilizer needs. This is the principle advantage of soil testing.

Don't guess when applying lime and/or fertilizer. There's a better way to manage soil fertility and pH than playing a kind of gardener's Wheel of Fortune. Just take soil samples from your lawn or garden and send them to your state laboratory or to a reputable private testing service. You will receive, in return, a detailed report describing your soil's pH and fertility status and indicating what measures need to be taken to grow the plants you desire.

How To Take Soil Samples

Collecting soil samples is not a difficult task, but it must be done properly to get the most meaningful results. Laboratory analyses are precise and accurate for the samples provided. So the better your sample, the better the results of the test. It is your responsibility to make sure the sample adequately represents your yard or garden.

Think of a "soil sample" as meaning the composite of many borings, cores or spade slices from one distinct area. To get a representative soil sample, gather soil from ten to twenty places chosen randomly throughout the area. Then combine and mix these to make a single sample for that location.

Individual samples should represent distinct areas or growing conditions based on present use, past treatment and/or future plans. For example, your yard could be divided into various sampling areas, such as lawns, flower beds, rose gardens, shrub borders and vegetable plots. The front lawn would be one sampling area and the back lawn another, particularly if one area was graded and the other not. Areas that have been limed and fertilized differently in the past should be sampled individually.

Depths of samplings should vary depending on tillage and current condi-

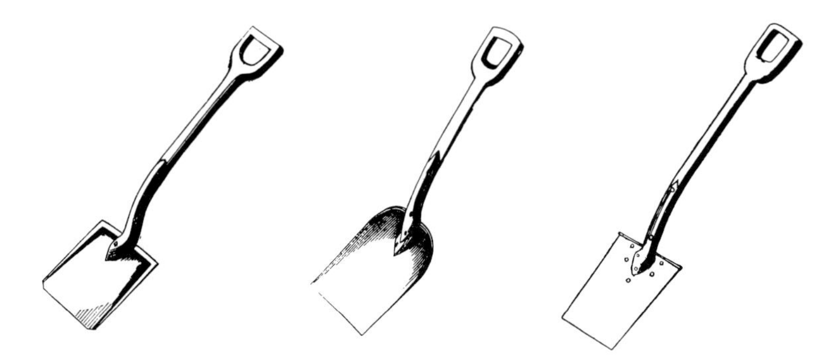

tions. For plantings that are already established and where no tillage is planned, sampling to a depth of two to three inches is preferred. Where tillage is planned or prior to plant establishment, or for areas routinely tilled, take the sample from the tillage depth.

Sampling Tools

Several tools can be used to collect soil samples:

Soil Sampling Tube. This is the most convenient. Simply push the tube to the depth to be sampled. Collect cores from various points and combine them.

Shovel and Tablespoon. Push a shovel six to seven inches deep into the soil. Tilt its handle forward to make an opening wide enough to permit you to scrape some soil from the wall of the hole. With a tablespoon, and starting at the depth to be sampled, scrape upward to the soil surface. Then remove the shovel and let the soil fall back into place. Combine soils from various points for the sample.

Shovel and Trowel. Dig a hole with a shovel, then take a thin slice from one wall. With a trowel, mark off a one-inch-wide strip down the middle of the slice on the shovel and push off the excess on both sides. Combine these one-inch strips for the sample.

For all sampling procedures, use clean tools and containers, mix the samples well, then deposit each combined sample in a soil box or bag for delivery to the laboratory. Most County Extension Service offices have information about sampling, along with boxes or bags for samples and directions for sending them to a testing laboratory.

Home Testing Kits

Do-it-yourself testing kits have been around for several years and are still available. The more inexpensive varieties usually provide only colorimetric (color comparison) tests which are, at best, crude approximations. These may give some indication as to whether or not a nutrient is present, but do not adequately define a soil's relative fertility level. Some of the more expensive kits are more reliable, but their initial cost and the cost for replenishment of chemicals (which must be done yearly) is prohibitive for the occasional testing undertaken by the average homeowner. The fees charged by professional laboratories range from none to a modest rate per sample. And laboratories have the advantage of research information for your particular locale. Therefore, home test kits may be interesting to use, but for more precise evaluation and recommendations, a state or reputable commercial laboratory is best.

SOIL TESTING LABORATORIES

OF THE 50 STATES AND SIX CANADIAN PROVINCES*

STATE OR PROVINCE	DEPARTMENT OR INSTITUTION PERFORMING TEST	ADDRESS	SHOULD APPLICATION BE MADE DIRECTLY, OR THROUGH COUNTY AGENT?
ALABAMA	Soil Testing Laboratory Auburn University	Auburn University, Alabama 36849	directly
ALASKA	Soil Testing Laboratory Agricultural Experiment Station, University of Alaska	533 E. Firewood Palmer, Alaska 99645	County Agent
ARIZONA	Soil, Water and Plant Tissue Testing Lab, Dept. of Soils, Water & Engineering University of Arizona	Tucson, Arizona 85721	County Agent
ARKANSAS	Soil Testing & Research Laboratory, University of Arkansas	Fayetteville, Arkansas 72701	County Agent
CALIFORNIA	No soil testing service is offered by a public agency.		
COLORADO	Soil Testing Laboratory Colorado State University	Fort Collins, Colorado 80523	either
CONNECTICUT	Soil Testing Laboratory Plant Science Department University of Connecticut	Storrs, Connecticut 06268	either
DELAWARE	Soil Testing Laboratory University of Delaware	Newark, Delaware 19711	County Agent
FLORIDA	Soil Testing Laboratory University of Florida	Gainesville, Florida 32611	County Agent
GEORGIA	Soil Testing & Plant Analysis Laboratory, University of Georgia	Athens, Georgia 30602	County Agent
HAWAII	Soil Testing Laboratory University of Hawaii	1910 East-West Rd. Honolulu, Hawaii 96822	either
IDAHO	Soil Testing Laboratory Dept. of Plant & Soil Science College of Agriculture	Moscow, Idaho 83843	County Agent
ILLINOIS	No soil testing service is offered by a public agency.		

** In addition to those listed here, private laboratories offer soil testing services.*

State or Province	Department or Institution Performing Test	Address	Should application be made directly, or through County Agent?
Indiana	Plant and Soil Analysis Laboratory Agronomy Department Purdue University	West Lafayette, Indiana 47907	either, preferably directly
Iowa	Soil Testing Laboratory Iowa State University	Ames, Iowa 50011	either
Kansas	Soil Testing Laboratory Agronomy Department Kansas State University	Manhattan, Kansas 66506	County Agent
Kentucky	Soil Testing Laboratory University of Kentucky	Lexington, Kentucky 40546	County Agent
Louisiana	Soil Testing Laboratory Department of Agronomy Louisiana State University	Baton Rouge, Louisiana 70803	County Agent
Maine	Soil Testing Laboratory Department of Plant and Soil Sciences University of Maine	Orono, Maine 04469	directly
Maryland	Soil Testing Laboratory Agronomy Department University of Maryland	College Park, Maryland 20742	County Agent
Massachusetts	Soil and Plant Tissue Laboratory University of Massachusetts	240 Beaver Street Waltham, Massachusetts 02254	either
Michigan	Soil Testing Laboratory Michigan State University	East Lansing, Michigan 48824	County Agent
Minnesota	Soil Testing Laboratory University of Minnesota	St. Paul, Minnesota 55108	directly
Mississippi	Soil Testing Laboratory Cooperative Extension Service Mississippi State University	Mississippi State, Mississippi 39762	County Agent
Missouri	Soil Testing Laboratory University of Missouri	Columbia, Missouri 65211	County Agent
Montana	Soil Testing Laboratory Plant and Soil Science Department Montana State University	Bozeman, Montana 59717	either
Nebraska	Soil Testing Laboratory University of Nebraska	Lincoln, Nebraska 68583	County Agent

State or Province	Department or Institution Performing Test	Address	Should Application Be Made Directly, or Through County Agent?
Nevada	Soil Testing Laboratory Plant Science Department University of Nevada	Reno, Nevada 89507	either
New Hampshire	Analytical Services Laboratory, University of New Hampshire	Durham, New Hampshire 03824	County Agent
New Jersey	Soil Testing Laboratory S & CS Department Rutgers University	New Brunswick, New Jersey 08903	County Agent
New Mexico	Soil & Water Testing Laboratory, Crop & Soil Science Department, New Mexico State University	Las Cruces, New Mexico 88003	either
New York	Soil Testing Laboratory Agronomy Department Cornell University	Ithaca, New York 14853	County Agent
North Carolina	Soil Testing Laboratory Agronomic Division North Carolina Department of Agriculture	Raleigh, North Carolina 27611	either
North Dakota	Soil Testing Laboratory North Dakota State University	Fargo, North Dakota 58105	directly
Ohio	Soil Testing Lab Ohio Res. & Dev. Center Ohio State University	Wooster, Ohio 44691	County Agent
Oklahoma	Soil Testing Laboratory Agronomy Department Oklahoma State University	Stillwater, Oklahoma 74078	County Agent
Oregon	Soil Testing Laboratory Oregon State University	Corvallis, Oregon 97331	directly
Pennsylvania	Soil Testing Laboratory College of Agriculture Pennsylvania State University	University Park, Pennsylvania 16802	County Agent
Rhode Island	Soil Testing Laboratory University of Rhode Island	Kingston, Rhode Island 02881	either
South Carolina	Soil Testing Laboratory Clemson University	Clemson, South Carolina 29634	County Agent
South Dakota	Soil Testing Laboratory South Dakota State University	Brookings, South Dakota 57007	directly

State or Province	Department or Institution Performing Test	Address	Should Application Be Made Directly, or Through County Agent?
Tennessee	Soil Testing Laboratory University of Tennessee	Nashville, Tennessee 37211	either
Texas	Soil Testing Laboratory Texas A & M University	College Station, Texas 77843	either
Utah	Soil Testing Laboratory Utah State University	Logan, Utah 84322	either
Vermont	Soil Testing Laboratory University of Vermont	Burlington, Vermont 05405	either
Virginia	Soil Testing Laboratory Agronomy Department Virginia Polytechnic Institute	Blacksburg, Virginia 24061	County Agent
Washington	Soil Testing Laboratory Washington State University	Pullman, Washington 99164	either
West Virginia	Soil Testing Laboratory West Virginia University	Morgantown, West Virginia 26506	County Agent
Wisconsin	Soil & Plant Analysis Laboratory, University of Wisconsin	511 Mineral Point Rd. Madison,Wisconsin 53705	either
Wyoming	Soil Testing Laboratory Plant Science Department University of Wyoming	Box 3354 Laramie, Wyoming 82071	either
Alberta	Soil and Feed Testing Laboratory University of Alberta	O.S. Longman Building, 6909 116 Street Edmonton, Alberta	directly
British Columbia	Soil Testing Unit British Columbia Department of Agriculture	1873 Spall Road Kelowna, B.C. V1Y 4R2	directly
Manitoba	Department of Soil Science University of Manitoba	Winnipeg, Manitoba R3T 2N2	either
Nova Scotia	Soils and Crops Branch Nove Scotia Agricultural College	Truro, Nova Scotia B2N 5E3	either
Ontario	Soil Testing Laboratory Department of Land Resource Science, Ontario Agricultural College University of Guelph	Guelph, Ontario	directly
Quebec	Canadian Industries Limited Soil Laboratory Beloeil Works	McMasterville, Quebec	directly

ABOUT LIME AND SULFUR

Changing Soil pH

John Paul Bowles

Lime and sulfur are the two primary soil additives used to alter pH—lime to raise it (make the soil more alkaline or "sweeter") and sulfur to lower it (make soil more acid).

All liming materials contain calcium, but all calcium-containing compounds are not lime. Gypsum, for example, is calcium sulfate. Gypsum is useful for making soils high in sodium fit for growing plants. It also adds calcium to calcium-deficient soils (calcium is an essential plant nutrient), but gypsum does not have much effect on pH.

Liming materials can be divided into two categories: man-made and naturally-occurring. Man-made limes include wood ashes (when the fire is deliberate), blast-furnace slag, hydrated lime and quick lime. Naturally-occurring limes are calcic limestone or calcium carbonate, dolomitic limestone (contains calcium carbonate and magnesium carbonate), ground mollusk shells and marl.

Which lime is best? Liming products are evaluated for their pH changing ability by comparing the effectiveness of each to 100-percent pure calcic limestone—calcic limestone has a calcium carbonate equivalent (CCE) of 100.

Hydrated lime and quick lime are caustic, to you and your plants, and more expensive than limestone. Each has a CCE of over 100, but their disadvantages outweigh their advantages.

Blast-furnace slag is rarely available in a form suitable for horticultural use and may contain undesirable by-products. Wood ashes are fine to use, but vary in liming ability depending on the type of wood burned (fresh wood from hardwood, deciduous tree species produces the best lime). It is interesting to note that wood ashes leached by water become lye.

Marl is a finely particled mineral deposit found at the sites of old, dried-up lakes and on existing lake bottoms. It is okay to use if you can find it. In some parts of the country lime from shells is available, but the shells must be very finely ground to be of any use.

Of the remaining liming products, calcic and dolomitic limestones are the cheapest, easiest, safest to use and the most readily available.

Dolomitic limestone contains about 20 percent magnesium carbonate, so it not only limes soil, but adds magnesium (another essential plant nutrient). Since it's almost impossible to overapply magnesium, it's always acceptable to use dolomitic limestone. In most of the soils of the U.S., however, magnesium is seldom lacking.

Calcic and dolomitic limestones' effectiveness in making soils more alkaline is dependent on the size of their particles. Remember, the two limes are naturally occurring *stones* and must be ground up for use. The smaller the limestone particles, the more rapidly soil reactions take place and pH change occurs.

Limestone is seldom pure, however, and can vary in its CCE. The CCE of useful limestone should be at least 80. Happily, modern agricultural and horticultural limestone is rarely sold in a form that has a CCE less than that figure.

How much to apply? That depends on soil type. The finer the texture of a soil, the more lime required to change its pH. Of the three particle components of soil—sand, silt and clay—clay particles are the smallest, sand the largest. Every soil particle has sites where ions can be attached, detached or exchanged as soil chemical reactions take place. The attachment of calcium ions to these exchange sites is what affects soil pH. Any equal volume of a clay soil versus that of a sandy soil has many, many more particles and possible reaction sites. Because of this, the clay soil requires much heavier lime applications to be affected toward a pH change.

Also, the difficulty of changing pH increases with an increase in the quantity of organic matter (OM) present. Organic matter is loaded with ion exchange sites. Organic matter, in part because of its high amounts of hydrogen and sulfur, tends to induce lower soil pH. Such things as the percentages of OM and clay in a soil affect the soil's buffering capacity—or its resistance to pH change. This is why it's not a simple "trick" to change pH.

However, many U.S. agricultural soils tend to be clay loams or loamy clays (fine to moderate in texture) with average or low amounts of OM (five percent or less) present in the upper six inches. For home gardens with these kinds of soils, about nine pounds of limestone (CCE=80) are needed to raise the pH one point on the pH scale. You may need to use more if your soil is unusually heavy or high in OM. There *is* one advantage to having heavy or fine-textured soils—because of their high buffer capacities, it is difficult to overapply lime.

The title of this article is "About Lime and Sulfur"—so what about sulfur? Although treatments with other materials besides sulfur will lower pH (many nitrogen fertilizers and some phosphorus fertilizers, for example), elemental sulfur is the best way to reduce pH and acidify soil. Elemental sulfur, sometimes called flowers of sulfur, is just sulfur—nothing else—powdered to make it react quicker. Aluminum sulfate is sold in small quantities as an acidifying agent, and it will lower pH. However, plants are vulnerable to aluminum toxicity and for this reason, aluminum sulfate should not be, or only rarely, used.

Just as for lime, the buffer capacity of a soil determines the effectiveness of sulfur. And sulfur and sulfur-containing acidifying fertilizers, such as ammonium sulfate, usually do not permanently (or rapidly) change pH. So tests and periodic re-applications are required.

Even though a soil test is the most accurate way to find out if sulfur is needed and how much to apply, a general rule of thumb is that two to three pounds of sulfur applied per 100 sq. ft. will lower pH 1.0 unit.

"Why bother with all this pH adjustment business?" you may ask. "After all, wild things seem to do okay." True, but then you don't see English oaks, Japanese hybrid azaleas or potatoes springing up spontaneously in your backyard either (if you do, then you don't need this handbook). The point is, the plants you grow are *not* native to the site you've chosen, and plants *are* pH sensitive. Some are more pH tolerant than others and grow in a wider range, but all have a spot on the pH scale that they like best. Most prefer a range between 5.5 and 6.5, although many are tolerant of levels as high as 7.2 or as low as 5.

Many nutrients become unavailable—or, worse, toxic—when pH is too high or too low. Also, desirable soil fungi and bacteria grow best in a pH between 6 and 7. The addition of lime to some soils improves soil structure by encouraging aggregation of particles, which, in turn, aids drainage and aeration.

Managing pH for specific crops can affect insect and disease control. For example, bluegrass grows well in neutral to alkaline soils, but Japanese beetle larvae *do not.* White potatoes will produce in a pH between 5 and 7, but potato scab *does not* like a pH below 5.5.

By now, I'm sure you'd like to know how to apply the materials. Both sulfur and lime can be surface-applied, but both are more effective and work faster if cultivated in. Neither should be applied to wet foliage and should be watered-in soon after application if in contact with dry foliage. Neither should be applied in windy weather (they are both dusty). Lime should not be applied at the same time as fertilizers. But lime applied long before fertilizer may increase fertilizer uptake. If pH needs to be altered more than one pH unit for soils with existing plants, then apply only enough lime or sulfur each spring or fall to change pH one unit or less until the desired level is achieved. Also, applications of either material may *continue* to affect pH one or many years later, especially if surface-applied.

Soil testing has been mentioned over and over again. If you feel that the recommendations from soil testing laboratories have been too general, then perhaps you have been too general in your submissions. If you are concerned about a specific crop, such as blueberries or azaleas, then send in a separate soil sample from the prospective planting site of each. Or, when submitting a general soil test, provide a list of the specific materials you're planning to plant — deciduous azaleas, apple trees, corn, for example, not flowering shrubs, vegetables and annuals. The soil testing laboratory, however, may not have all the answers, so keep developing that green thumb (and keep a file of pH preferences for your favorite plants) so that one day you'll be able to do the "by-the-seat-of-your-pants" kind of estimating that will make you a real "gardener."

REPRINTED, WITH PERMISSION, FROM *THE DAWES ARBORETUM NEWSLETTER.*

Adding lime to soil raises the pH, making it more alkaline or "sweeter." Many nutrients in the soil become unavailable — or worse, toxic — when pH is too high or low.

PHOTO COURTESY USDA SOIL CONSERVATION SERVICE

Suggested Applications of Powdered Sulfur to Lower pH

Figures are for application when combined with cultivation. If applying only to the surface, reduce figures by two-thirds and apply no more often than once a month in warm seasons—less in cold months. For heavy soils, increase recommendations by one-third. Use only powdered or pelleted elemental sulfur, not iron sulfate (at least, not often) and do *not* use aluminum sulfate.

Approximate Lbs. of Sulfur per 100 Sq. Ft. of Soil Tilled to a Depth of 8 In.

Desired ph		4.5		5.0		5.5		6.0		6.5	
Soils		A	B	A	B	A	B	A	B	A	B
	5.0	3/4	2 1/2								
Existing	5.5	1 1/2	4 3/4	3/4	2 1/2						
pH	6.0	2 1/2	6 3/4	1 1/2	4 3/4	3/4	2 1/2				
	6.5	3	9 1/2	2 1/2	6 3/4	1 1/2	4 3/4	3/4	2 1/2		
	7.0	3 1/2	12	3	9 1/2	2 1/2	6 3/4	1 1/2	4 3/4	3/4	2 1/2

A=Sandy Soil B=Loam Soil

General Recommendations for the Amount of Finely Ground Limestone Required to Raise pH

Lime moves downward, but not laterally, in soil. Spread uniformly over the surface. Mix with soil wherever possible to hasten effect. It is preferable to use no more than 50 lbs. of ground limestone per 1,000 sq. ft. per application. If more is required, space applications several weeks apart (depending on rainfall or irrigation).

Pounds of Limestone per 1,000 Sq. Ft. of Lawn Area.

	Existing Soil pH	Sandy Soils	Sandy Loams	Loams & Silt Loams	Clay Loams & Clayey Soils
very	4.0	90	120	165	200
acid	4.5	80	105	150	180
	5.0	70	90	120	150
	5.5	45	60	90	120
	6.0	25	30	45	60
neutral	7				

SOIL CONDITIONERS

NANCY HOWARD AGNEW

When garden soils are inadequate for sustaining vigorous plant growth, soil conditioners may be used to improve tilth. If used properly, conditioners improve the aeration, drainage and friability of tight or heavy garden soils, and the moisture-retaining ability, without water-logging, of light, sandy varieties. Soil conditioners actually modify the structure and texture of soil to achieve these results.

Soil conditioners may be organic (such as peats and composts), mineral (such as sand) or chemical (lime or gypsum). Organic and chemical conditioners function differently, but both help aggregate soil particles and improve physical structure. Mineral conditioners modify soil texture (or the distribution of soil particle size) to improve aeration and drainage.

Organic Conditioners

Organic conditioners are the more versatile of the two kinds. Immediately after application of an organic conditioner, soil becomes more aerated because of the spongy nature of the raw material. The porous organic material also increases soil water-holding capacity. As organic conditioners decompose, a biological glue or cement that binds soil particles together into aggregates is created. The better aggregated the soil particles are, the better the soil's aeration and drainage. Organic materials provide improved plant nutrition because they tend to prevent leaching of soluble fertilizers, and because nutrients are released through bacterial-action decomposition of the conditioners.

Peats

The majority of commercial organic conditioners are comprised of peat. Several kinds of peat can be used as conditioners, but not all of them are suitable for every situation. Peats vary in pH, water-holding capacity and degree of decomposition; these characteristics determine whether or not a peat is appropriate for a particular situation.

SPHAGNUM-MOSS PEAT, a fair soil conditioner, is light brown in color. It has a fast rate of decomposition, which makes it a poor choice if the addition of stabilized organic matter is desired. Sphagnum-moss peat is oligotrophic, meaning it is low in nutrients (only 0.6 to 1.4 percent nitrogen). But it is excellent for use with

NANCY HOWARD AGNEW *is temporary Assistant Professor of Horticulture, Iowa State University, Ames, Iowa.*

ericaceous (acid-loving) plants because it has a pH range of 3.0 to 4.0. It is also lightweight and has excellent water-holding capabilities.

HYPNUM-MOSS PEAT, a good soil conditioner, is medium to dark brown in color. Its moderate rate of decomposition makes it fair as a stabilized organic-matter addition. Hypnum-moss peat is eutrophic, meaning it is rich in nutrients (2.0 to 3.5 percent nitrogen), and with a pH range of 5.0 to 7.0, it is a good soil amendment for non-ericaceous plants.

REED-SEDGE PEAT, partially decomposed reeds, sedges, marsh grasses and cattails, is usually a good soil conditioner, but it can be quite variable in pH and degree of decomposition. Low-lime reed-sedge peat has a pH range of 4.0 to 5.0 and contains 1.5 to 3.0 percent nitrogen. Overall, reed-sedge peats have a slow rate of decomposition and are a good source of stabilized organic matter; they have medium water-holding capacity.

PEAT HUMUS, dark brown to black in color, is a good soil conditioner. It is highly decomposed and is an excellent source of stabilized organic matter. Peat humus has a pH range of 5.0 to 7.5 and contains 2.0 to 3.5 percent nitrogen. Because of its advanced state of decomposition, peat humus is not spongy and has low water-holding capacity.

Manures and Composts

Green manures (or cover crops) are excellent organic soil conditioners. Green manures, however, take six months to one year to grow. Once mature, they are turned under the soil. Then enough time for decomposition to take place must be allowed before the benefits of soil conditioning can be reaped. Cover crops are less expensive than peat because the cost of seed is comparatively low. Annual rye and buckwheat (sown at a rate of two to three pounds per 1,000 sq. ft.) and rape (sown at a rate of two to five ounces per 1,000 sq. ft.) are all good green-manure crops.

Animal manures can also be soil conditioners. Cow and horse manures are best. Fresh chicken and sheep varieties must be used in small quantities because of plant-burn potential. For this reason, they are not as desirable for soil conditioning. For best results in improving soil texture and adding slow-release nutrients, manure should be composted before use.

Garden compost makes a good soil conditioner. It is inexpensive, but takes time to produce. Its value as a source of nutrients, as a soil amendment for increasing water-holding capacity and as a source of stabilized organic material varies with the stage of decomposition and type of parent material.

Mineral Conditioners

Mineral conditioners improve drainage in tight or heavy soils, such as clay. The primary mineral conditioner used to modify garden soils is sand. It improves drainage by creating larger pores, but the wrong sand or quantity of sand can clog pores instead of enlarging or increasing them, with cementlike soil the result. Usually, additions of stabilized organic matter and perhaps chemical conditioners along with good soil management (don't till when wet, avoid compaction, etc.) are more effective than mineral conditioners for improving heavy soils.

For soil amendment, washed coarse silica sand of the brick or mason type is preferred. Particle diameter should range from 0.25 to 1.0 mm., or the sand should be described as medium- to coarse-textured.

Other mineral conditioners include calcined clay, blast-furnace slag or expanded shale; all should be selected in the particle-size range indicated for coarse sand. Vermiculite and perlite are not recommended because they lack mechanical strength; they will become crushed in garden soils and will tend to impede rather than enhance drainage.

Chemical Conditioners

Chemical soil conditioners include lime (calcium carbonate) and gypsum (calcium sulfate). These compounds, rich in calcium ions, flocculate (bring together) dispersed soil particles into larger particles. When soil particles are dispersed, water is unable to penetrate, much in the way a drop of water is unable to penetrate talcum powder. Soils high in sodium (sodic soils) tend to be greatly dispersed. These soils are generally located in arid regions and in areas irrigated with water high in sodium. Lime and gypsum flocculate sodic soils by exchanging calcium ions for sodium ions. When soil particles flocculate and form aggregates, aeration and drainage is improved. Chemical conditioners are most effective in soils with this unique condition and are not useful for "loosening" compacted soils or soils with high clay content. In non-sodic soils, lime and gypsum are applied primarily to affect pH or add nutrients.

Cover crops or "green manures," such as rape in the foreground above, are excellent organic soil conditioners.

Leaves are composted in the homemade bins above, providing an excellent source of organic matter and nutrients for the garden.

MANURES AND COMPOSTS

MARY LEWNES ALBRECHT

Earlier generations knew that if you returned to the earth part of what you took from it, it would continue to produce bountiful harvests. Native American Indians taught the pilgrims to place a fish at the bottom of planting holes; they knew that the crop would grow better that way. Farmers use tractors and manure spreaders to return animal wastes to the fields that produced the animals' feed. Home gardeners put such practices to use in flower and vegetable plots.

The incorporation of animal

MARY LEWNES ALBRECHT *is Assistant Professor of Floriculture, Kansas State University.*

Average Nutrient Percentages of Some Manures & Organic Fertilizers

All figures are per 100 pounds

	Nitrogen (N)	Phosphorus (P)	Potassium (K)
Fresh Cow Manure, Including Bedding	.5 lbs.	.3 lbs.	.5 lbs.
Dried Poultry Manure without Litter	4.0 lbs.	3.0 lbs.	3.0 lbs.
Dried Rabbit Manure without Bedding	2.4 lbs.	.6 lbs.	.05 lbs.
Fresh Sheep Manure	1.0 lbs.	.4 lbs.	.2 lbs.
Fresh Horse Manure	.4 lbs.	.2 lbs.	.4 lbs.
Blood Meal	15.0 lbs.	1.3 lbs.	.7 lbs.
Bone Meal	4.0 lbs.	21.0 lbs.	.2 lbs.
Cottonseed Meal	3.2 lbs.	1.3 lbs.	1.2 lbs.
Farrowing Swine Manure	4.0 lbs.	1.0 lbs.	negligible

The composting of fresh manures often lowers their nutrient content because of leaching; however, well managed composting can significantly raise nutrient levels, as initial water-to-solid-material ratio is reduced (fresh manure contains a large amount of water).

manures, composts and green manures into garden soils improves soil structure and workability and adds nutrients. Unfortunately, these organic sources do not supply large quantities of the macronutrients — nitrogen, phosphorus and potassium — that are most essential for plant growth and development. They are, however, good sources of essential micronutrients such as copper, zinc, molybdenum, boron, iron, manganese and chlorine.

Animal Manures

Livestock operations must deal with waste disposal. In feedlots water is used to wash manure from feeding pens. Quite often, this is stored in an effluent pond that is drained on a regular basis, cleaned out and the deposits spread on agricultural land. With stabled animals, bedding is spread to absorb urine and moisture from manure. The used bedding is either spread in fields or stocked in large composting piles. The nutrient content of manure varies considerably depending upon animal species and diet. Feedlot animals tend to produce richer manure than those that are pasture-grazed. Also, pasture-grazed livestock eat "weeds" and, of course, weed seeds. Many native plants found in pasturelands (weeds if in a garden) require a special scarification (the breaking of the seed coat) prior to germination. For many weed seeds, the digestive tracts of grazing livestock provide an excellent scarification treatment. But passed undigested with manure, the seeds, ready to grow, may be added to garden beds. This may also be a problem with the manure of stabled animals which are often pasture-grazed, as well as feed-fed. Proper composting procedures kill most of the weed seeds in manure.

Types of Animal Manure

Poultry manure, commonly available in

Manure? Compost? Green Manure?

What's the difference between manure, compost and green manure? Manure is the excreted waste of animals; it may be fresh or composted, processed or unprocessed, mixed with straw or other material used as animal bedding, or undiluted. Compost is any material that has begun to decompose, and it is usually ready for the garden near the end of its decomposition process. It is often wholly derived from plant residues, but it can contain manure or even *be* manure.

Compost is preferred to fresh material because dead plant parts must decay for nutrients to be released, and the decay process itself uses nutrients. But after a certain point in the decomposition process, nutrients are released and made available for living plants. Also, benefits to soil structure are derived only from decomposed plant residues (if you add undecomposed leaves to a soil, the leaves must decay before they have a beneficial effect). And fresh manures contain chemicals that can be damaging to plants. Green manures are living plants used to improve soil fertility and structure. They also prevent erosion in soils that are not in production.

bags at garden centers, varies considerably according to the type of poultry it originates from—layers versus broilers, for example. Nutrient content is also affected if the manure has been hydrolyzed (steamed under 30 pounds of pressure for 30 minutes, then passed through a drier and finally into a hammer mill) or dried (dehydrated to a moisture content of 10 to 15 percent and passed over a one-quarter-inch sieve). Dried poultry waste, which has a greater nutritive value than the hydrolyzed variety, has an approximate analysis of 4-3-3, with less than one percent magnesium and about nine percent calcium. It also contains measurable amounts of manganese, iron, copper and zinc. Dried poultry manure makes an excellent soil amendment when worked in at the rate of 10 to 20 pounds of manure per 100 square feet of garden bed.

Swine manure, from farrowing yards, is normally mixed with straw bedding. Nutrient analysis is about four percent total nitrogen and one percent phosphorus, with very little potassium. Zinc and iron are also present at trace levels. Wood ashes or muriate of potash (potassium chloride) are possibilities for supplemental potassium. The major drawback of swine bedding as a soil additive is the copper fed to pigs to promote growth and discourage bacteria. The copper is concentrated in the pigs' urine and absorbed in the bedding. If swine manure is used often, copper can accumulate in soils and cause plant toxicity.

Other types of animal manures commonly added to garden soils are cattle, horse and sheep. In *fresh* form, these manures range from about one percent nitrogen for sheep manure to less than one-half percent for that produced by cattle. Phosphate content, at four-tenths of a percent per volume, is highest in sheep manure, and even less for cattle manure. Potassium content for either is

A variety of composting techniques are demonstrated at the Tilth Community Garden in Seattle.

less than one percent. *Dried* cattle and sheep manures have a higher nutritional analysis because of the loss of water and subsequent concentration of nutrients. Nitrogen, though, can be lost due to volatilization, and if compost piles are kept too wet, additional nutrient losses are caused as a result of leaching.

Horse manure is considered to be "hotter" since its nitrogen is more readily available to plants and its nutrient value is short-term. The values of cattle and pig manures are longer lasting, due to a higher content of more slowly available nitrogen. The use of straw or sawdust bedding is desirable because it absorbs much of the moisture from manure and urine and helps prevent nutrient loss.

Use of Manures and Composts

Manures can be handled in two ways: they can be tilled into the ground when fresh, or added to the compost pile to improve it. But fresh manure cannot be safely used around either new or established plants. Commercially dried and bagged manures can be used directly in the garden during spring preparation, or even later in the season.

If using fresh manures, till them into the top six or eight inches of the soil in the fall. This allows the manure to break down into safe forms during the winter. If young plants are set into soil that has had fresh manures incorporated, plant damage can occur as a result of ammonia volatilization. Fresh manures should not be used in bulb beds for the same reason. Spread fall applications of fresh cattle and horse manures at the rate of 50 to 100 pounds per 100 square feet; spread fresh sheep manure at the rate of 10 to 20 pounds per 100 square feet.

The best manures to use as soil addi-

Compost can be made in simple heaps, foreground left, bins or drums, and should be worked into the soil on an annual basis.

tives are those that have been composted with sawdust, straw, leaves and/or other plant wastes. Plant residues that make excellent additions to a compost pile include grass clippings, autumn leaves, remains of flower and vegetable plants, weeds that have not gone to seed, peelings and seed pods. Never add disease-or insect-infested material to a compost pile. The addition of dry vegetative matter helps absorb nutrients, preventing their loss. Keeping the manures moist reduces the volatilization losses that normally occur during composting.

An advantage of adding manures to composting plant residues is the additional nutrient compounds they supply. These are necessary for microorganism growth (which is necessary for composting to occur). Thus, the addition of manufactured fertilizers to the compost pile is not necessary if manure is used. But most animal manures have low potassium content. Wood ashes are a rich source of potassium (five to 10 percent by weight). However, because wood ashes weigh so little, considerable amounts must be used.

Tilling manures into the ground improves the structure and workability of soils—better aeration and drainage in heavy clay soils, better "gluing", or particle-clumping action in sandy soils. The latter augments micropore or capillary spacing, making more water available to plants.

Manures or composts must be incorporated as soil amendments on an annual basis. In warm, humid climates organic matter decomposes quickly in the soil and there is little humus accumulation. In drier, cooler climates there may be accumulation, but annual additions are still necessary to improve problem soils; regular additions of composts or manures contribute micronutrients and help maintain good soil structure.

Composts of animal and vegetable wastes can be used as mulch. A three- to four-inch layer spread over the soil helps prevent moisture loss from evaporation. But in certain parts of the country, organic mulches encourage increased sowbug populations. Normally, sowbugs feed on dead or dying plant tissue. They pose a problem in the garden when their populations increase to the point that they begin feeding on live material. Periodically, the soil under the mulch should be examined for sowbugs. If large populations are present, control measures, such as insecticide application or reduced irrigation (if possible), should be implemented.

Another commonly used organic mulch is uncomposted grass clippings. But in some areas grass clippings ought to be buried in the compost pile for complete decomposition prior to garden use. Fresh grass clippings are a favorite site for stable-fly breeding, a serious problem in rural areas; unlike house flies, stable flies have a nasty bite. Also, fresh grass clippings tend to mat together as they dry, becoming impermeable to water.

Green Manures

Another technique long used to improve soil structure is green manuring. Green manures are special crops that are grown in fields or beds needing improvement. Green-manure plant seeds are generally sown in the fall and turned under in the spring. This operation not only offers the benefit of returning nutrients to the soil and adding organic matter, but the bare ground is protected against wind and water erosion. So green manures also function as cover crops. Green manures are a soil improvement alternative when animal manures are in limited supply.

Green manures are always plants with high water content that decompose rapidly. As green manures break down in soil, there is a release of carbon dioxide and an increase in organic acids which results in temporary acidification, making some unavailable forms of essential nutrients and some micro-nutrients accessible.

Nitrogen is contributed if leguminous cover crops are used. Legumes, such as peas or alfalfa, form symbiotic relationships with certain bacteria species that fix (or acquire) atmospheric nitrogen. The bacteria form nodules (or small swellings) on the roots of the legumes. In these nodules, the bacteria convert atmospheric nitrogen into ammonia which is then utilized by both the bacteria and the plant. When the cover crop is tilled into the ground, a portion of this fixed nitrogen becomes available for subsequent crops.

Common cover crops or green manures include annual rye grass (one to two pounds of seed sown per 1,000 square feet), rye (three to four pounds per 1,000 square feet), oats (two to three pounds per 1,000 square feet) and wheat (three to four pounds per 1,000 square feet). Other non-leguminous green manures include barley, mustard, Sudan grass, millet and buckwheat.

Red clover, alfalfa, sweet clover and cowpeas are used as legume green manures. Check with your local extension office to find out which are winter-hardy in your area. All cover crops should be tilled into the ground early in the spring or prior to seed set.

Regardless of the types of manures or composts available, the time and effort put into them result in more rewarding garden harvests. With the continued use of any manure or compost, soils become lighter, have improved water-holding capacity and aeration and provide a better environment for plant growth.

MAKING "BLACK GOLD"

One Woman's Prize-Winning Method

Marilyn D. Walker

What's dead and alive, good for nothing and useful for everything, worthless and valuable, repulsive and desirable, malodorous and fragrant, wasteful and conserving?—Compost!

Paradoxical as it seems, a person making compost takes "worthless," often malodorous waste and by creating conditions conducive to microbial life, joins with Mother Nature in returning nutrients to the soil.

Ever since green plants began to grow on land there has been death and the recycling of nutrients. Vegetation that dies and falls to the ground nourishes millions of bacteria and fungi, which in their turn die, providing nutrients for plants which evenutally die, etc., etc. If this happens naturally, why go to the trouble of making compost? There are five good reasons:

1) Finished compost is a good substitute for manure. In modern times and crowded places there is not enough manure to use as a soil supplement in food production.

2) Bacteria that decompose dead vegetation require a lot of nitrogen. When vegetation is left to rot on its own, the bacteria seize all the soil nitrogen available and keep that nitrogen from growing plants until the decomposition process is finished. (But, of course, in the end the bacteria die and release the nitrogen to the soil and to plants.)

3) Composting speeds up the natural process to as short a time as fourteen days by some methods.

4) Weed seeds and plant pathogens can be killed by the heat generated by composting if the pile is properly made.

5) If only for the sake of good housekeeping around yard and garden, composting is a worthy endeavor because it takes crop residues, manure and other wastes, and gives back to the earth rich, fragrant "black gold."

Making compost is its own reward. There are many workable methods. Each gardener eventually picks his/her own favored techniques according to location, climate and materials available.

My Own Pile

I have a large — approximately 10,000 square feet — garden in Ohio. Here is my method for making sandwich-layer compost:

Marilyn D. Walker *is coordinator of the summer gardening program for Lancaster, Ohio, city schools. She is an herbalist, homestead gardener and educator.*

These compost bins are tucked into a corner of a garden in Northern California.

Starting with a five- by five-foot square, I lay down a 10- to 12-inch-deep base (layer one) of twigs, cones or very coarse material, so there is aeration from below. Then I add a 6-inch-deep cover (layer two) of crop residues, weeds and kitchen garbage (no meat scraps or greasy residue).

I keep the layers relatively flat, or slightly concave in the middle, and squared out to the corners. This is essential for water absorption (moisture is vital to composting); the pile should be as wet as a wrung-out sponge, and every layer needs sprinkling before the next is added.

After the second layer is placed and moistened, a two-inch layer of fresh manure is added. The manure is usually already moist and contains straw bedding. Some compost "recipes" call for a layer of soil, but I think there is soil enough on the weed and crop roots. Soil in the compost pile is necessary because it helps keep nitrogen from leaching. The three layers should come to a total thickness of about two feet.

Atop the third layer, I place several two-inch-diameter PVC pipes horizontally. The PVC pipes have one-inch-diameter holes cut in them every few inches to provide aeration to the pile (like water, air is essential). Make sure the pipes are long enough to stick out the sides of the finished heap. The pipes free me from having to turn the pile. (A cubic yard of compost weighs one ton! Who wants to turn that inside out?) Then I continue layering, as above, until the pile reaches four feet or so in height; if the pile were any higher it would be too weighed down, preventing good aeration.

In my region we can have very rainy weather between November and May, so to prevent nutrients from leaching I usually cover the pile with black plastic through the winter. Also, in hot, dry weather a cover prevents loss of moisture. I place carpets over the plastic and sometimes add weighs at the corners to keep them from blowing up. In an

unsheltered winter location, a windbreak of straw bales is a good idea. Compost made in spring and summer is ready in three months; fall and winter piles are normally ready in the spring.

The principles that my technique follows are:

1) a balanced blend of carbonaceous and nitrogenous materials; the ration is three to one, respectively, or about a six-inch layer to a two-inch layer.

Carbonaceous material, or vegetation containing lots of carbon, is supplied by dry, coarse material: twigs, the straw in the manure and tougher parts of other "stuff" added to the pile. Nitrogenous, or nitrogen-containing, matter is supplied by the manure and soft parts of the plant

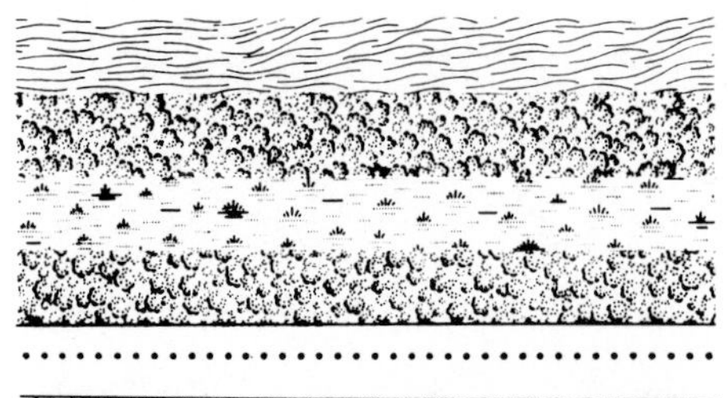

materials. In general, the greener, softer or more succulent (water-containing) a plant part is, the more nitrogenous it is, and vice versa.

The desirable carbon/nitrogen (C/N) ratio is based on the quantity of carbon and nitrogen needed by decomposition bacteria (which also just happens to be the ratio of carbon and nitrogen making up the bacteria). Many things commonly added to compost piles—dry leaves and straw, for example—have a very large C/N ratio. The C/N ratio of straw is about 120 to one. Twigs would, of course, have an even greater C/N separation. The C/N ratio needed by decomposition bacteria is closer to 20 or 30 to one, so a compost pile containing only straw decomposes slowly and inefficiently unless additional nitrogen is supplied. However, fibrous material (dry tree leaves as opposed to green lawn clippings) produces more substantial compost.

I've tried making compost without manure, and it did not heat up enough to kill the weeds. High nitrogen additives can also take the form of sprinklings of fish, blood, cottonseed or seaweed meal, or even commercial fertilizer. Don't be reluctant to add extra nitrogen to a compost pile—in composting (like so many other gardening activities) you get back as much (more?) as you put in.

2) The provision of good aeration for the growth and survival of decomposition bacteria. Compost piles without aeration promote anaerobic (functioning without oxygen) bacteria that smell bad, cannot digest organic material completely and produce some substances that are toxic to plants.

3) the addition of the right amount of moisture required for microbial activity. A too-dry pile takes too long to break down. Note: a foul odor can mean the pile is too wet.

4) The construction of a pile large enough to "heat up," stay hot and compost. The minimum size seems to be at least three feet square by three feet tall (a cubic yard). Maximum height is about five feet.

So why not give composting a try?. You'll find that it is easy, efficient and effective.

MULCH
A COVER STORY

MICHAEL E. ECKER

Mulching was invented by nature eons ago. A forest floor with many thin layers of leaf litter in various stages of decomposition is mulching at its best. Successful mulching is really quite simple and is literally down to earth.

What is mulch? Mulch is any material covering soil. Mulch serves a number of purposes: it controls erosion, conserves water, moderates soil temperature and prevents weed growth; sometimes it's just used for decorative purposes.

MICHAEL E. ECKER *is Assistant Horticulturist at The Dawes Arboretum, Newark, Ohio.*

Adding mulch to the soil surface of flower beds, vegetable gardens or under and around trees and shrubs has *many* advantages: soil retains moisture longer (an especially important factor in sandy soils); existing weeds are inhibited from growing and their seeds are kept from sprouting; soil temperature is moderated, keeping soil cooler in summer and warmer in winter; rapid freezing and thawing that can "heave" small plants out of the ground is avoided; mowers are kept a safe distance from plants and mowing is eased by rounding out difficult corners; soil erosion and compaction from rain and foot traffic is

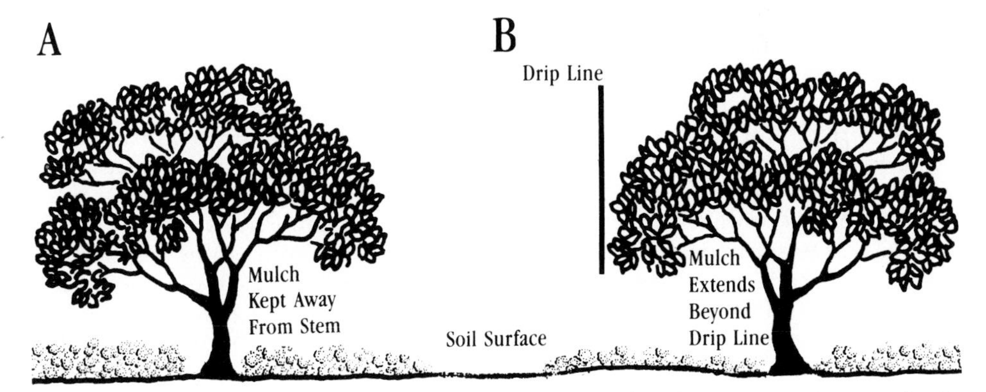

Coarse-textured mulch, such as wood chips or bark (A), can be applied deeper than fine-textured mulch, such as sawdust or grass clippings (B). The more of the root system covered, the more beneficial the mulch.

Mulch controls erosion, conserves water, moderates soil temperature and prevents weed growth — and adds a decorative touch.

prevented; fruits, leaves and flowers stay cleaner (also walks, walls, low windows, etc.) and landscapes are made more attractive, with contrasting areas visually unified.

Organic mulches are materials that were once alive (woodchips, leaves, etc.); inorganic mulches are plastic, stone and the like. Organic mulches have advantages over inorganic; nutrients are recycled and soil structure is improved as decomposition progresses. Also, organic mulches can be worked into the soil if desired, whereas it is necessary to remove inorganic kinds if an area is to be cultivated.

Another reason for adding mulch is to protect against herbicide injury. Some pre-emergent herbicides (which stop the germination of seeds) volatilize readily in warm weather. Unless a protective mulch barrier is placed over a treated area, smaller plants may be injured or killed by the released gasses.

If treating green weeds or grasses with contact herbicides, mulch may be applied first or afterwards; however, slightly better weed control is achieved if the herbicide is applied a few days before the mulch.

There are many types of organic mulch: wood chips, shredded bark, bark chunks, sawdust, sphagnum peat moss, crushed corn cobs, pine needles, grass clippings, peanut hulls, leaves, straw, manure containing straw, newspaper (don't worry about lead—inks no longer contain it), spent hops, cocoa bean and buckwheat hulls, compost, hay; the possibilities are limited only by imagination and accessibility.

The material used should be airy, allowing adequate air circulation for optimum growing conditions (because of

SOIL AERIFICATION BY CORING
WHAT IS IT, WHAT IS IT GOOD FOR?

Compacted or heavy soils have poor water percolation, drainage and air exchange. When an established lawn is growing in such a soil it is desirable to try to improve growing conditions through aerification. This is a process by which small plugs, or cores, are removed from the soil. Coring provides for better air movement in and out of soil, increased water penetration and a decrease in thatch.

Coring is best done in early fall on cool-season lawns and in early summer for warm-season lawns. Cores can be removed in a variety of ways but a motor-driven corer is probably the best. Material removed by coring should be broken up and spread back over the surface of the lawn. Devices that simply make holes in the ground (usually called spike aerators) instead of removing plugs are not recommended.

this, sawdust is not recommended). It should also be durable, slow to burn, easy to handle and attractive if used ornamentally. Any organic mulch must be replenished occasionally because organic matter decomposes. Inorganic mulches last indefinitely, but many types eventually disintegrate (paper, plastic), especially if exposed to light.

If fresh sawdust, corn cobs or other fine-textured fresh organic mulches are used, additional nitrogen must be

The more airy a mulch the deeper it can be applied. Wood chips or bark should be applied about two or three inches deep. Six to eight inches of straw or pine needles is recommended.

applied because decomposing bacteria utilize nitrogen that would otherwise be available to plants. For a mulch to decompose, soil microorganisms must feed upon it. These microorganisms require oxygen, water and nitrogen. Plants exhibit stunted growth and a yellow color if, because of competition with decomposition organisms, nitrogen deficiency becomes a problem. When applying fertilizer to correct this situation, use a complete analysis fertilizer with a 1:1:1 ratio, such as a 10:10:10. Pull existing mulch to the side and mix the fertilizer into the top two inches of soil. Mulches decompose more rapidly if they are mixed with fertilizer.

Be cautious when using finely shredded mulches such as peat, sawdust or bark. Once wetted and then dried, the top layer can form a sort of pie crust, impeding water penetration. The development of certain fungi that sometimes grow on mulch, especially chips and shredded bark, can create a water barrier also.

Leaves are a satisfactory mulch because they are widely available, provide trace elements through decomposition and attract earthworms. Leaves present a few problems, though. They have a tendency to blow from where they're placed, and when wet they mat or pack down, which is good for weed inhibition but also inhibits water penetration. The problem can be minimized somewhat by

Mulch Across and Down

Mulching and the Improvement of Compacted Soil for Plant Health and Winter Tolerance

It can (and usually does) get very cold in much of the United States. Even though most of the plants we grow are adapted to winter's chill, many (particularly young, recently transplanted or older plants under stress) benefit from extra protection.

A three- to six-inch-deep layer of *coarse* mulch such as partially rotted leaves, composted grass clippings, woodchips, bark or even straw spread over a plant's root zone can be compared to a heavy down vest that not only keeps the plant warmer by holding the earth's heat, but lessens the possibility of damage from drought and sudden changes in temperature. Mulching not only helps relieve stress in winter, but also helps remedy problems that result from the sun and thirst of summer.

The root zones of established plants extend beyond their drip line, or foliage spread. For young established plants it is advisable to mulch an area one-third to one-half again as large as the area covered by the plant canopy. For older and larger plants at least one-half the canopy-shaded root zone should be mulched. Organic mulch can be spread anytime, but it's best to apply it after temperatures are regularly 45 degrees and lower. Fertilizer, to spur root development and spring growth, can be spread before mulching, at the rate of five pounds of actual nitrogen per 1,000 square feet of soil surface.

Physiological stress caused by compacted, poorly drained and/or drought-subjected soils is common; compacted soil is especially prevalent in urban and suburban landscapes. Many landscape plants such as flowering dog-

chopping the leaves into smaller pieces and/or mixing with other materials such as grass clippings or pine needles.

Keep thick organic mulches from coming into contact with the bases of trees and shrubs. Rodents working under mulch cover can damage or even kill young plants. Fungi and slugs that injure plants find a warm, moist, dark area (such as under mulch) a haven. If mulch is applied too deeply, overly wet conditions may result. The more airy a mulch is, the deeper it can be applied. For example, when applying wood chips or bark, two to three inches is desirable, whereas a depth of six to eight inches of straw or pine needles is recommended.

If a large area is to be covered and the mulch is delivered via heavy equipment (i.e., heavier than a wheelbarrow), have it dumped near, but not on, the site to avoid compacting the soil. A wheelbarrow may take longer but is far more considerate of soil structure.

Many inorganic or synthetic materials are suitable for more permanent mulches; choices include: black plastic (polyethylene), woven, or fiber weed, mats (polypropylene), aluminum foil or asphalt paper. Crushed stone, or pebbles, is more ornamental but does little to inhibit weeds or retain soil moisture. If used, however, mineral aggregates should be contained by some type of edging, or they will be scattered and lost. Rocks are suitable as an

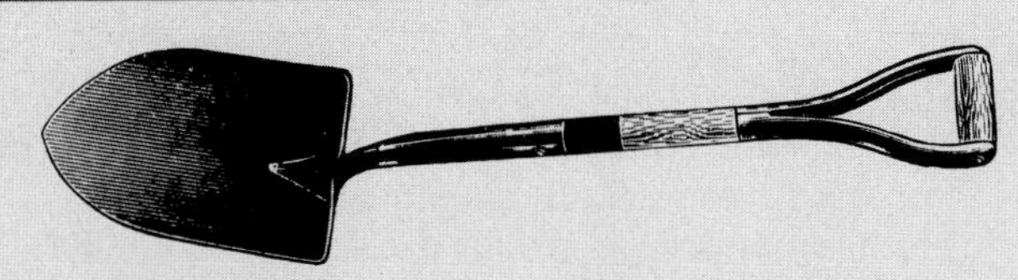

wood and sugar maple are particularly intolerant of compacted soils.

An aid for this problem has recently been developed. The procedure, known as vertical mulching, is an aerating process that allows air and water to drain into and flow out heavy and compacted soils. To mulch plants vertically, drill holes one or two inches in diameter and 18 inches deep on 12- to 18-inch centers. Position the holes in concentric circles throughout the root zone of the plant subject (remember, roots extend far beyond the drip line). Don't place holes near the plant's trunk—most feeder roots are nearer the drip line, and holes close to the trunk may damage large, supporting roots. Holes one inch in diameter do not need to be filled, but larger holes should be stuffed with a one-to-one mixture of sphagnum peat and fine gravel (coarse sand, calcined clay [e.g., Turface] or shale [e.g., Haydite grade S-1] may be substituted for the gravel). Or you can use one of the coarse mineral materials alone.

If you like, horizontal mulch of a coarse organic material can be applied on the soil surface after vertical mulching, but vertical mulching is effective even in turfgrass-covered areas. Like traditional mulching, this new kind of mulching can be done every year, and should be done several years for severely compacted soils. Timing is not critical; vertical mulch any time the soil is not wet or frozen.

Incidentally, very young and recently transplanted trees should have burlap wrap applied. The burlap should extend from close to the ground up past the first branches. This will reduce rabbit and deer damage and help prevent frost cracks and sunscald.

Organic mulches have advantages over inorganic mulches like plastic and stone: Their nutrients are recycled and soil structure improves as they decompose.

ornamental covering over other permanent mulches, however.

The great disadvantages of using plastic film or any other nonporous mulch is prevention of evaporation and difficulty of irrigation. Plant roots may suffocate beneath the mulch from too much water, or dry up, even in rainy periods. If holes are punched to alleviate these conditions, then weeds become a problem. Also, nothing is more hideous in the landscape than poorly anchored, flapping black plastic.

In vegetable and flower gardens, root growth will be better if mulch is applied in late spring, after soil has warmed up. If planting trees or shrubs in the fall, mulch right away to keep soil around the plants' roots warmer longer, so they continue to grow.

The type of mulch used depends on where an area is located in the landscape. Strawy composted manure, a bit smelly but a good mulch nonetheless, should not (obviously) be used around entranceways or patios. Mulches that are somewhat flammable, such as straw and leaves, should not be applied wherever cigarettes might be tossed. If it's winter mulch protection that's desired, looks should be secondary to function. But deep winter protective mulches (for insulating tender perennials, semi-evergreens and bulbs) that are applied thicker than normally recommended must be removed in the spring, before growth begins.

One further note: you cannot mulch too large an area beneath a tree (at least as far as the tree is concerned). Make the mulched area as extensive as is practical, even past a tree's drip line, if possible. Not only is it good for the tree, but it also means there is less lawn to mow.

HOW TO APPLY WATER

ROBERT N. CARROW

PHOTO BY ELVIN MCDONALD

The amount of water needed for plant growth relies on two major considerations: a) evaporation losses to the atmosphere from moist soil and plant surfaces, and b) transpiration losses. Transpiration is a process that helps plants take up water. During transpiration a plant absorbs soil moisture through its root system and then "transpires" most of it (95 percent or more) through openings (stomata) in the leaves. Untranspired water is retained for plant growth.

ROBERT N. CARROW *is Associate Professor of Agronomy, University of Georgia.*

The two water-use components combined are called *evapotranspiration* (ET). ET rates vary tremendously, depending on plant species, air and soil temperatures and amount of wind and sunlight. Two other means of water loss are runoff and leaching below the root system. The goal of good irrigation management is to replace the quantity of water lost by evapotranspiration, while eliminating runoff and leaching losses.

Conditions that stimulate ET, reduce the quantity of ET or lower the amount of water available to a plant's roots (such as low soil water-holding capacity or

Table 1
Plant and Atmospheric Factors that Influence Plants' Water Use

Plant

1. *Species* — Species vary in their water requirements. Kentucky bluegrass, for example, requires more irrigation than tall fescues or fine-leaf fescues, while bermuda grass and zoysia use less water than St. Augustine grass.
2. *Rooting Depth* — Good physical and chemical soil conditions will allow maximum root growth for a particular species. Proper management such as fertilization and mowing also promotes good rooting. A deep-rooted plant requires less irrigation.
3. *Stomatal Control* — When a plant wilts, the stomata close and prevent further water loss by transpiration. Unfortunately, the plant will decline in vigor and may die if this stress persists.

Atmospheric

1. *Temperature* — High temperatures increase plant water use.
2. *Solar Radiation* — Bright, sunny days stimulate evapotranspiration.
3. *Humidity* — Plants grown in arid or semi-arid climates use more water than the same plants grown in a humid climate.
4. *Wind* — Winds up to about four mph increase water requirements but have little influence beyond that rate. However, high winds distort irrigation patterns and prevent uniform applications of water.

shallow rooting) result in the need for frequent irrigation. Some plant and atmospheric factors that influence water use are listed in Table 1. Soil physical conditions also affect irrigation practices and will be discussed later.

To develop a proper irrigation program, you will need to know the answers to the following questions:

How Much Water Should I Apply?

After a rainfall or irrigation, the total quantity of water available to a plant depends on: a) the water-holding capacity of the soil and b) the depth of the plant's root system. Soils differ in their ability to hold water for plant use (see Table 2). For example, a turfgrass with a one-foot deep root system growing in a sandy loam soil would have .9 to 1.3 inches of water available for ET before drought stress occurs; once irrigation is necessary, between .9 to 1.3 inches of water would have to be added to the soil. Thus, knowledge of the soil's water-holding capacity is a guide to how much must be added to replenish the water used in evapotranspiration.

A second factor to consider is the depth of the plant's root system. For example, a turfgrass with a two-foot deep root system has twice the available water as a similar plant with a one-foot system. The deeper rooted grass requires *less*

TABLE 2
AVERAGE WATER-HOLDING CAPACITIES AND INFILTRATION RATES FOR DIFFERENT SOIL TEXTURE CLASSES

SOIL TEXTURE	PLANT-AVAILABLE WATER PER FOOT OF SOIL* (INCHES OF WATER)	INFILTRATION RATE** (INCHES OF WATER PER HOUR)
Sand	0.4-1.0	.50-8.0
Sandy Loam	0.9-1.3	.40-2.6
Loam	1.3-2.0	.08-1.0
Silt Loam	2.0-2.3	.06-0.8
Clay Loam	1.8-2.1	.04-0.6
Clay	1.8-1.9	.01-0.1

* Soil also holds some water too tightly for plants to extract. The "plant-available water" in this Table is that water potentially available for plant extraction.

** These values vary widely depending on surface conditions (compaction, temperature, etc.).

frequent irrigation, but when irrigation is necessary, *more* water is needed to replenish soil moisture.

Gardeners must evaluate rooting depths and adjust required irrigation depths on a seasonal basis for most plants. A dormant plant requires much less water than it does when actively growing. Some plants, such as lawn grasses and ornamental perennials, have greater or lesser root development from season to season, depending on their periods of active growth.

A good rule of thumb for watering the garden: Irrigate to a level slightly deeper than the plant's root system. To evaluate how much water to apply, use the information in Table 2 at the top of the page; apply the estimated quantity of water when irrigation is required; check the depth of water penetration after 24 hours; adjust future irrigation quantities based on how deeply the irrigation replenished soil moisture. Soil probes that will help you judge rooting depth and water penetration are available at garden centers or mail-order houses.

How Often Should I Irrigate?

Standard rules of thumb on frequency of irrigation, such as, "irrigate once per week using one inch of water," are usually not very accurate. Atmospheric and plant conditions affecting water use (see Table 1) vary substantially over the growing season. Thus, frequency of irrigation should be adjusted throughout the season, with more irrigation in the summer.

"Indicator" spots or plants may help in timing irrigations. Most home lawns, for example, have areas of grass that exhibit the first indications of moisture stress, with other areas showing signs one to two days later. By observing indicator spots, gardeners are provided with guides to help schedule the irrigation of specific sites.

Shaded locations often need less frequent irrigation, unless severe root competition from adjacent trees or shrubs is a problem.

Plants often need water during dormant periods. Irrigation may be required during the winter if long periods of above-freezing temperatures occur or if the winter is dry, with little snow cover. This is a significant problem in the Central Plains region.

How Should I Apply Water?

Growers should know how much (i.e., the *rate*) water is applied per hour by their particular irrigation equipment. Rotary systems usually apply 0.25 to 0.45 inch per hour, while spray systems apply 0.75 to 1.0 inch per hour. To check a system, set empty coffee cans every 10 to 15 feet in a grid pattern in the zone watered. After irrigating for one hour, measure the depth of water in each container. Figure the average by adding the depths in all the containers and dividing by the number of containers.

With this information, you can determine how long a site must be irrigated to achieve a desired quantity of water. As an example, if 1.5 inches of water is planned for application to a lawn, and the irrigation equipment used delivers 0.75 inch per hour, two hours of operation would be required.

If the soil has a high percolation of infiltration rate, water can be applied in one setting of a sprinkler. However, many times the soil's infiltration rate (see Table 2) is *less* than the amount the irrigation system delivers. If so, an attempt to apply all the water in one continuous setting will result in excessive runoff. One approach is to apply 50 to 70 percent of the water in one setting, wait for two to 24 hours and then apply the remaining water. This allows the water more time to infiltrate the soil, and eliminates runoff. This method is useful on heavy (high clay content) soils, compacted soils, sloped areas and turfgrasses with excessive thatch. Infiltration rates can often be increased by cultivation or other mechanical methods of aeration.

Besides the rate of water applied, the *uniformity* of an irrigation system is important. When collecting water in coffee cans, you may note that some cans contain very little. This will certainly be evident if you use only one sprinkler at a time to irrigate. Identifying areas that receive too little or too much water helps eliminate dry or wet spots.

In-ground irrigation systems (those buried in the soil) should be designed so they apply water uniformly. If one or two spots receive less water than needed, while remaining areas are adequately irrigated, then those sites must be spot-watered with an end-of-hose sprinkler. Using a water *breaker*, or rose, is the best way to hand irrigate. Do *not* overwater an entire area just to irrigate dry spots adequately.

To achieve uniform irrigation, single sprinklers should be moved about with sufficient overlap. As a rule, reset sprinklers at a distance 50 to 55 percent of the diameter they irrigate (with a wind of greater than four mph, use a 40 to 45 percent spacing). To illustrate, a single sprinkler that covers a circle with a 75-foot diameter should be moved to a new position that's 38 to 41 feet away (assuming low wind conditions).

What Time of Day Should I Irrigate?

The best time to water is very early in the morning when the wind is calm, city water pressure is good, evaporation losses are minimal and leaves remain moist for only a few hours, until the sun dries them (this helps prevent foliage disease).

Another good time to irrigate is in the early evening. Some people are concerned that allowing leaves to remain moist overnight may promote disease. This is not a problem when a site is irrigated all in one evening, or two consecutive evenings, and then several days are allowed to pass before the next irrigation. Under such a regimen, you do not need to worry about disease development.

The least efficient time to irrigate is in the afternoon when winds are highest, water pressure is lowest and evaporation losses are high. Some individuals believe that afternoon watering causes "burning" of the leaves because water droplets focus and intensify the light. This is not true, but if a turf or garden area is overwatered to the point that the soil is saturated and standing water is evident — and temperatures are very high — the plants can "scald." Under these conditions plant roots will be deprived of the soil oxygen they need to remain healthy. They won't be able to absorb water, even when submersed in it, because of lack of oxygen. And they may die from the subsequent prevention of transpirational cooling.

What Kind of Equipment Should I Use?

End-of-hose sprinklers are the most common home irrigators. These are relatively inexpensive but require the labor of having to be moved frequently. The one most often used is the rotary type that emits water in one or two directions at a time, while slowly rotating. A second kind is the sprinkler that emits a fan of water, while waving back and forth over an area. A third is the variety that sprays water in a circle in all directions continuously. The last applies water over a smaller area, but at a high application rate. All these configurations are available with part-circle or part-wave adjustments.

Another system worth considering is in-ground piping with quick-couplers. Either rotary or spray sprinklers can be connected to the couplers. When several sprinklers are employed at one time, this system is more flexible and much easier to use than the single, end-of-hose arrangement.

The most flexible system is the automatic kind in which an electric or mechanical timer activates irrigation of an area by segments (zones). Each zone is controlled by an automatic valve that feeds water to several sprinklers (rotary, spray or wave). When the zone valve is activated by the controller clock, these sprinklers begin irrigation. The clock/timer can be programmed so that a zone is watered at specific times of the day and only on the days indicated.

Automatic systems must be carefully designed so that heads are properly spaced, and, if possible, sprinkler application rates are at or below those of soil infiltration. While an automatic system is the most costly of all, it is a very efficient means of irrigation if properly designed and programmed.

In short, efficient irrigation depends on uniformly applying the correct *quantity* of water at the proper *time* interval and at the appropriate *rate*.

PHOTO BY ELVIN MCDONALD

PLANTS ARE NOT ANIMALS

Understanding Plants' Needs

John Paul Bowles

Nitrogen, along with at least sixteen other elements, is essential for healthy plant growth. Along with nitrogen, hydrogen, oxygen, carbon, phosphorus and potassium are all universally required for life, not just by plants, but by animals too.

For plants, calcium, magnesium, sulfur, iron, boron, manganese, copper, zinc, molybdenum and chlorine complete the essential sixteen. Perhaps you noted that most of these elements are also required by humans, which seems to imply that there is more similarity between plants and animals than is superficially discernible. But there is one (among many) important difference: animals eat, plants do not.

Animals can only receive complete nourishment by breaking down (through digestion) very complex compounds (food) into the simpler forms needed as physiological and anatomical building blocks. Plants, however, can only make use of compounds and elements that are already in a simple form and similar to or exactly the same as the construction material needed for biological growth. Actually, the nutrient bits that plants initially absorb in soil water or air are already simpler than most compounds derived through digestion in animals.

Eating (and breathing, too) is an activity enjoyed only by animals; plants cannot go looking for their equivalent of a hamburger when the need arises. A plant must make do with what is in its immediate vicinity — in the air and water that comes in contact with the leaves, stems, and roots.

For those readers with probing minds who have said, "Wait a minute! What about Venus-flytraps?", consider this: a Venus-flytrap may trap flies as a nutritional supplement, but it doesn't chew, can't swallow or taste, is entirely nonselective (a piece of inert cellophane gets the same treatment as a juicy insect), cannot move on if the "hunting" is poor and can (and often does) live quite nicely, thank you, with out its buggy snacks if plenty of the right elements are available in adjacent soil water. No, plants do not eat, and the nutrients plants get from air — carbon, hydrogen, oxygen and some nitrogen — are obtained through their roots, and drift to their vicinity in the same way as all the other nutrients plants absorb.

Notice that in reference to plants, the words, food and vitamins, have not been used. That's because the only correct way to refer to the sixteen essential elements (and any other elements or compounds required for plant growth) is plant nutrients. It has already been established that plants do not eat, so it follows that they cannot take in food. The word vitamin refers specifically to compounds needed for animal (originally only human) growth and development. Vitamins needed for animals may sometimes come from plants, but the application of the word and the physiological application of the substance has no parallel in plants.

Fertilizer is sometimes called plant food, even plant vitamins, but now you know this is wrong, right? Fertilizer refers to any material put on or in soil to encourage plant growth. Since fertilizer must stimulate or improve plant growth, then it must contain one or more plant nutrients. (A material that improves plant growth only by improving soil structure is a soil conditioner.) Fertilizer comes in several forms: animal manure, compost or as a pelleted, granular or powdered processed product.

Plants may not eat, but this doesn't mean they are not selective about the nutrients they use. Selective is probably not the best word because I don't mean to imply that plants can make choices. Phosphorus, for example, is one of the most common mineral elements in the earth's crust, yet it is a major fertilizer component. Why? Because phosphorus in many soils is either insoluble (won't dissolve in water) or is in a form that plants cannot use. Plants get the bulk of their nutrients from water taken in through their roots. If soil phosphorus is in an insoluble form, then plants cannot get at it. If the phosphorus in soil water is in a form plants can't absorb, then it just won't be used. Plants cannot change the phosphorus into the kind needed. And phosphorus, like most of the elements required by plants, comes in many forms, as well as parts of other compounds.

Nitrogen is one of the most common elements on earth (air is mostly nitrogen) and, like phosphorus, is a common fertilizer component. Unlike phosphorus, however, nitrogen, although highly transient, can be used in many forms by plants. Nitrogen in soil leaches readily, volatilizes back into the atmosphere or (you've heard this before) is in a form plants can't use.

But people have determined scientifically in recent years — and by trial and error in past ages — that composts and certain ingredients in prepared fertilizers supply nitrogen and phosphorus in forms plants can, and do, use.

So for plants to take up most nutrients, the nutrients must be in forms the plants can use and in soil solution. Soil solution is what soil water is called when referring to its nutrient properties. Soil is vital to plants for three reasons: 1) it provides support for the above-ground parts; 2) it is a source of nutrients, and 3) it is the medium through which the soil solution comes in contact with plant roots. Nutrients get into soil water in various ways: rocks are broken down by weathering (freezing, thawing, erosion); many complex soil chemical reactions take place that dissolve large compounds; rain carries nutrients from the atmosphere; microorganisms, through their particular modes of living (and dying), greatly contribute to the availability of essential nutrients, and we contribute by spreading fertilizer on soil surfaces and setting out the sprinkler.

REPRINTED, WITH PERMISSION, FROM *THE DAWES ARBORETUM NEWSLETTER*.

PLANT NUTRIENTS
Macro, Micro, Major Minor and Lbs. per Sq. Ft.

JOHN PAUL BOWLES

Plant nutrients are grouped according to the quantity required and the necessity of the nutrient as a result of scarcity. Macronutrients are those that are needed in relatively large amounts. Nitrogen, phosphorus, potassium, calcium, magnesium and sulfur are macronutrients. Because available nitrogen (N), phosphorus (P) and potassium (K) are commonly in short supply in most soils, they are known as major or primary nutrients (the "Big Three" found in most fertilizers). The other macronutrients—calcium, magnesium and sulfur—are called secondary nutrients.

Micronutrients (often referred to as minor elements) are all other nutrients needed for plant growth. These include iron, boron, manganese, copper, zinc, molybdenum and chlorine.

Sometimes in certain soils, secondary nutrients or even micronutrients are in such short supply that they become major nutrients, but this is fairly uncommon. (There are exceptions: iron, for example. However, the application of calcium, magnesium or sulfur to affect pH is not considered fertilization because the application is not for nutritive benefits.)

Fertilizer recommendations are usually based on nitrogen supply because nitrogen is typically the most crucial nutrient in shortage. Also, over the years, commercial fertilizers have come to be pronounced in N:P:K ratios that are suitable for most horticultural needs. And the ratios do vary somewhat across the U.S., depending on regional soils and crops. Plant fertilizer requirements or recommendations are usually given in X pounds of nitrogen to be applied to an area. So if recommendations indicate a fertilizer application of 10 lbs. of nitrogen per 1,000 square feet, then the approximate quantities of phosphorus and potassium needed are usually made available, too. This is not only convenient, but correct, since the amount of one nutrient in soil water affects the availability of other nutrients; too much of one nutrient can cause a shortage of another, just because soil-solution balance is off. For example, a complete fertilizer (one that includes all of the big three) with a ratio of, say, 18:5:9, generally supplies the necessary amounts of phosphorus and potassium for the plant group intended (trees and shrubs) if the nitrogen application rate is calculated and used properly.

Happily, manures and composts also tend to supply phosphorus and potassium in agreeable quantities, although a lot more materials must be used. For example, there are, on the average, 10 lbs. of nitrogen, 2.7 lbs. of phosphorus and 7.5 lbs. of potassium per *ton* of dairy cow manure; but in *100 lbs.* of commercial fertilizer 18:5:9 (including base materials—inert matter such as certain types of clays or waxes used as carriers or fillers) there are 18 lbs of nitrogen, 5 lbs. of phosphorus and 9 lbs. of potassium. So 100 lbs. of 18:5:9 has about 36 times more nitrogen than the same amount of cow manure. Of course, as stated in other articles, manures, unlike typical commercial fertilizers, almost always improve soil structure.

REPRINTED, WITH PERMISSION, FROM *THE DAWES ARBORETUM NEWSLETTER.*

TILLING AND SOIL IMPROVEMENT

JOHN PAUL BOWLES

Issues that are as hotly debated as politics (and almost as subjective), are how and when to till. These decisions are largely a matter of opinion. You can dig a hole for a plant or plow a garden plot any time or under any circumstances if you are aware of and willing to accept the consequences. For example, you might plant in soil that was tilled when too wet; almost everybody has (including me). But later in the season, because of poor aeration and root growth in the compacted soil, more irrigation will be required.

Wait a minute! Tilling can cause compaction? Of course it can, and an easier-to-visualize example is the case of the broken water line. Remember last year, or the year before, when that pipe in the back yard leaked and had to be repaired? Isn't there still a sunken spot, even though the same amount of dirt went back into the hole as came out? Digging a hole is a kind of tilling. If you till when soil is too wet, it will compact as it settles. Cultivated soil should be friable — moist, but crumbly. Tilling when soil is too wet can cause a condition known as puddling. The soil structure of puddled soil is so disturbed that percolation of water is essentially non-existent.

Okay, I know what you are thinking, and you are right. You *can* begin to correct this kind of compacted soil by tilling again when soil moisture levels are lower. But in central Ohio (where I live) and just about everywhere else, at least in spots, soils are naturally "heavy" (fine textured and clayey) and/or have been so drastically disturbed (or the plants you want to grow need such exceptional conditions) that a good tilling won't help. So it is best to till only when circumstances are just right. Tilling when soil is too wet is not the only way to cause it to become compacted. Overtilling, even when soil is

friable, causes compaction because desirable peds are crushed. Properly tilled soil should be coarse-grainy, not powdery or pulverized.

Typical garden cropping whereby plants are grown only for a few months and then mostly or completely removed, causes a reduction in organic matter content. The weight of equipment (even feet) and the force of implements used to cultivate may make a subsurface zone of compaction. The reason many farmers till, plant and fertilize all at once is not just to save time but to cut down on compaction caused by heavy equipment. Your feet crossing the lawn week after week, year after year, behind a mower compacts soil, particularly if you mow after a rain, if the stand of grass is poor (and little organic material is being accumulated in the root zone) or if you collect clippings (which can contribute organic matter to topsoil). Sandy soils are much less prone to compaction than loam or clay soils, but sandy soils usually tend to drain so readily that the addition of organic matter is necessary to increase water retention.

Rotary cultivators often not only pulverize a soil undesirably, but create a compacted zone below the levels tilled. Turning the soil with a spading fork, then raking it smooth is much more conducive to good structure than using a rotary cultivator.

Deep-digging is a procedure that is especially beneficial to the development of soils for finely rooted plants, such as vegetables and perennials. Deep-digging also helps improve drainage in heavy soils and in soils with compacted sub-surface layers, or pans. In deep-digging, the topsoil layer (normally six to 12 inches), is removed and set aside. Then sub-surface soil is spaded up and organic material is incorporated. Topsoil is then replaced and organic matter is incorporated into it, too. For deep-digging to be most effective, the total depth worked should be about two feet for most soils. Deep-digging, assuming it is done when soil moisture levels are low and the soil isn't overcultivated, is effective to a certain extent even if organic matter isn't added.

Tilling is really a way of managing soil so plants can be more easily started, grown and harvested. Another management technique is the addition of organic matter or other materials to improve soil structure, consistency and tilth. With all soils — clay, loam or sand — organic matter additions are the best (often the only) way to improve their physical nature.

Organic matter promotes aggregation of soil separates into peds which allow for increased percolation, drainage and water retention. Organic matter also has a stabilizing effect on pH. Another reliably beneficial soil additive and pH adjuster is lime. But lime is best added only when soil pH is too low (acidic) for good plant growth. It is very difficult to add advantageous quantities of sand to heavy soils. Even if enough were added

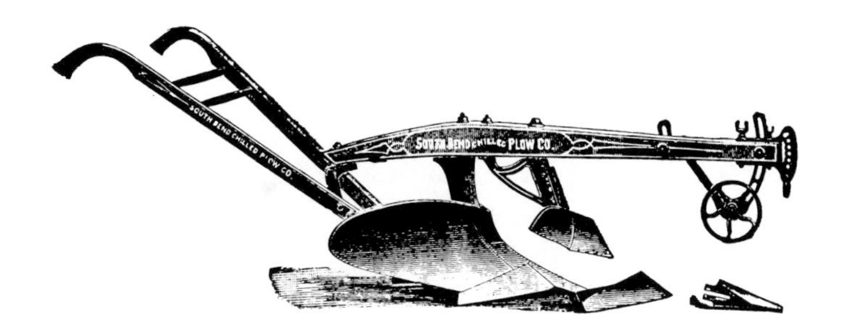

to a clay soil to make it a sandy clay, unless pH and organic levels were good, aggregation would remain poor.

Soil organic content is measured by weight. In fine-textured or heavy soils, five to 30 percent organic content is considered optimum — in coarse soils, two to 20 percent. For a clearer understanding of quantities, compare one pound of peat moss and one pound of clay and take note of the very large differences in volume. (A soil test can tell you the percentage of organic matter in a soil. Be sure to collect samples as you would for any soil test.) Contents of organic matter greater than 20 to 30 percent characterize soils that are called *organic* (versus *mineral*). Although good for plant growth initially, organic soils are difficult to stabilize because their structure changes as decomposition occurs.

How is organic matter content maintained in mineral soils?

1: Grow plants in an intensive arrangement and disturb as little as possible. Examples: stands of trees and shrubs in a "natural," woodslike setting, well managed lawns (clippings not removed). 2: Make surface applications of organic materials to established plantings. Example: mulching around a shrub border and replacing as needed. Rainfall, soil reactions and soil organisms incorporate the organic material into the soil. 3: Apply surface additions of organic materials to necessarily disturbed beds. Example: mulch in a flower border or vegetable garden for weed and water control and then incorporating the mulch when preparing the bed for the next planting. 4: Incorporate organic material. Example: spread a layer of compost on a garden and then till.

Which of the last three methods is best? It depends on when, how much and how often organic matter is applied. Fall is a better time than spring for incorporating organic matter because this allows a period of time for soil reactions to occur and for soil structure to improve and stabilize before plants are grown. However, because tilling exposes more soil to air, it causes increased decomposition and lower levels of organic matter. If a garden is cultivated for weed control, organic levels will drop for the same reason: increased exposure to air and heat lead to increased rates of decomposition.

The nature of the garden itself affects organic matter levels; a standard vegetable plot with rows of plants with broad spaces between row has a much lower organic content than an intensely planted plot as the ratio of root mass to soil mass is lower. Cutting at ground level, rather than pulling them up, results in higher organic matter levels.

Just as in a political race, the choice is yours. And as in politics, it is not the informed that make the wrong choices but the uninformed. An informed gardener thinks before he digs, and then digs with conviction.

Reprinted, with permission, from *The Dawes Arboretum Newsletter.*

MANAGING COMPACTED AND HEAVY (CLAYEY) SOILS

W. LEE DANIELS

Soils that are compacted or high in clay resist root, water and air penetration and can seriously hinder plant growth. Compacted soils are very dense and therefore lacking in pore space, which lessens available-water holding capacity and rooting area. Soils high in clay also tend to be dense. Because of the very fine texture and small pore size of clayey soils, water is so tightly held that uptake by plant roots is limited. Due to the lack of large pore spaces, water passes through both soil types very slowly; therefore, both are frequently too wet.

Management of Dense Soils

The ability of a growing root tip to penetrate soil is directly dependent on soil strength. Soil strength is controlled jointly by a soil's bulk density and moisture content. Workable, loose soils have bulk densities of 0.8 to 1.3 grams per cubic centimeter (or have 70 to 50 percent pore space), while severely compacted soils commonly have bulk densities in the range of 1.6 to 1.8 (or 40 to 30 percent pore space). Root penetration is greatly retarded when bulk density exceeds 1.4 during dry conditions. The same soil when moist, however, may not impede rooting because soil strength is then decreased.

Bulk density is difficult for the layman to determine accurately so, unfortunately, analysis of compacted or dense soils must be estimated. You can identify a clay soil (if you can't just by looking at it) by measuring its plasticity. To do this, take a small amount of soil from the area or layer in question and wet it. Roll the sample between your palms or on a flat surface. A clay soil makes a long thin ribbon, or string, that doesn't fall apart easily. The longer and more cohesive the "string" is, the more clayey the soil sample. Compacted soils or soil layers are (not surprisingly) hard to dig through, whether wet or dry. But a better way to determine compaction is to evaluate plant root growth. Plant roots will stop or be restricted in compacted layers, or root growth will generally be poor, as compared to growth in looser soils.

W. LEE DANIELS *is an instructor in soil science at Virginia Polytechnic Institute and State University. He is a specialist in surface mine reclamation and soil genesis.*

The best way to improve rooting in compacted soils is to increase the porosity by tillage and through the incorporation of organic matter into the soil. Addition of compost and/or other organic amendments into top-soil insures that soil is well aggregated (or better structured) and, therefore, contains more large pores, as well as total pore space.

Compacted layers or zones in soils are called pans. Compacted subsoil layers limit total soil volume available for rooting and restrict total water and nutrient availability. These layers also perch water tables (cause water to "back up" instead of draining downward). Sometimes, water tables perched by compacted soil layers last for extended periods of time. This causes saturation, or waterlogging, within the rooting zone. Deep tillage or ripping is the only practical way to improve subsoil porosity, but may be too expensive for many gardening situations. However, care must be taken to avoid excessive tillage, since that may lead to the destruction of large aggregates. Too much tillage also decreases organic matter content by speeding decomposition.

The structure of the compacted soil zone also strongly influences its affect on plant growth. Artificially compacted zones (traffic pans) are often much denser than natural clay pans, and have no breaks or channels, such as are found in naturally occurring pans. Traffic pans must be physically shattered in order for them to allow significant rooting. Natural subsoil clay pans, on the other hand, often have structural planes of weakness that allow roots to penetrate to some extent, even though the layer overall is

Soils that are compacted or high in clay, like the soil above, make it difficult for water, air and plant roots to penetrate and can seriously hinder plant growth.

PHOTO COURTESY USDA SOIL CONSERVATION SERVICE

very dense. Subsoil pans, especially naturally occurring ones, are permanent, frequently quite thick and are very difficult to alter.

Management of Wet Soils

Compacted and/or clayey soils cause numerous watering problems. The most obvious is surface flooding caused by slow water penetration into the ground. When downward water movement is limited by dense or high-clay layers, soil becomes saturated and oxygen (which moves very slowly through water) is kept away from plant roots. If the saturated condition persists, roots will die from oxygen starvation. Highly compacted soils, even when dry, cause the same problem. Extended periods of water saturation also lead to increased availability of heavy metals such as iron and manganese, which in some soils may actually poison plants. Saturated conditions accelerate soil nitrogen losses, particularly in cool climates and seasons.

There are a number of ways to manage saturation problems in soil. One is to increase internal water movement by improving aggregation and pore space. There are several ways to do this: increasing and maintaining organic material levels; changing or keeping pH in the range between 5.5 and 6.5; adding a soil conditioner such as very coarse sand; cultivating only when moisture levels are ideal, and avoiding compaction. But adding organic material is probably the single most effective action you can take.

Another way to increase internal water movement in wet soils is to shatter subsoil pans. If just a few deep passages for water are made down through the soil, large amounts of water will flow through them (assuming the underlying layers will accept the water). Or subsurface drainage can be installed beneath the soil to carry away excess water. This is usually expensive, but may be the only alternative in many situations. Still another approach is to limit the amount of water entering the soil by diverting surface water away from the poorly drained area, or by digging interceptor trenches just uphill from it. Plastic mulch can also be used to decrease total water penetration.

The best way to avoid saturation problems, however, is to learn how to recognize sites and soils that are prone to them. First of all, look at their position in the landscape. Sites that lie low, particularly at the base of long slopes, receive considerable amounts of subsurface water flow, and are almost always wet. Large flat areas with little surface drainage are also likely to have subsurface water problems, regardless of how high or low the large flat areas are. Gently sloping areas are always the best-drained, without being dry. Take care not to alter surface drainage patterns to insure that storm run-off doesn't flood the site. But this is often already a problem on many suburban and urban sites where natural water drainage has been changed by construction. Areas with dense subsoil pans or bedrock-perched water should be avoided whenever possible. Home building and other construction activities often result in severe compaction or the exposure of clay subsoil material.

A number of soil properties are indicative of both poor drainage and water saturation; even a novice soil scientist (that's you) can learn to recognize them. Look at the soil horizon, or the way a soil is layered. Both permanent and fluctuating water tables turn soil gray. Solid gray colors down a soil horizon indicate almost permanent saturation. Fluctuating saturation will cause mottled gray within the background soil color. Free-draining soil will have bright brown, yellow or red in their subsoils. ❦

IMPROVING SANDY SOILS

MANAGING SOIL AND WATER IN ARID REGIONS

FRED D. WIDMOYER

Very few garden soils are exactly as we would like them. Some are too sandy, or coarse, in texture. Sandy soils are easy to work wet or dry, and they warm up early in the spring. But they dry out quickly and often do not supply growing plants with sufficient nutrients, particularly nitrogen.

The addition of organic matter helps sandy soil retain moisture and nutrients longer. Many materials can be used to improve the water- and nutrient-holding capacities of sandy soils: sawdust, hay, bark, straw, cornstalks — with or without coarse-composted materials from the garden. If any of these is dry, hard and/or under-decomposed, apply nitrogen-containing fertilizer to help break it down. Work both the plant material and fertilizer into the soil thoroughly, so that they're evenly distributed.

If you use fresh sawdust as a sandy-soil amendment, add three to four lbs. of nitrogen (11 lbs. of ammonium nitrate or 17 lbs. of ammonium sulfate) to each cubic yard of sawdust (300 sq. ft., one inch deep) to prevent nitrogen deficiency. If smaller amounts of organic matter are applied, figure about 3/4 lb. of ammonium sulfate per bushel (32 quarts or eight gallons or 1.25 cubic feet) of material. If you use coarser material than sawdust, apply less nitrogen. Or use composted sawdust or other composted materials; pre-composted organic soil amendments don't require additions of nitrogen.

Watch plant development closely when using large quantities of organic matter. Slow growth, with small, pale-green or yellowish leaves, usually indicates that the plant is deficient in nitrogen and that a nitrogen fertilizer supplement is needed.

Generally, fresh animal manures and sewage sludge are not recommended for arid lands because they tend to increase saline (salty) conditions. However, if adequate irrigation is possible, no permanent plant damage will occur.

Arid Regions

Soil in low rainfall areas are often unfavorably alkaline. Soil pH can be lowered by increasing the amounts of peat moss, sawdust or ground-up bark, along with acidifying fertilizers. These organic materials tend to hold moisture and fertilizers in light-textured soils. Deep watering helps to remove alkaline minerals that may be a problem. However, soil that is only slightly alkaline, or nearly neutral,

FRED D. WIDMOYER *recently retired as Professor of Horticulture, New Mexico State University.*

will grow most garden plants.

Gardeners in regions of low rainfall may achieve greater success by the use of raised beds; it is much easier to modify soil in limited areas rather than entire sites. And drainage is improved when beds are built above existing grades. In some cases you can even grow plants requiring very acid soil by this method if beds are heavily amended to adjust pH levels.

Salty Soils

Excess salts occur naturally in most of the arid parts of the West. Salts are mineral compounds that may become so concentrated in soil water that nutrients, and water itself, become unavailable to plants. In some cases the salts occur naturally, or they may come from irrigation water, fertilizers or chemical applications. Water from shallow wells usually has a higher alkaline (mineral-salt) concentration than water from deep wells. Gardeners with salty soils must depend upon heavy leaching (applying great amounts of water periodically, especially before planting) to move salt concentrations to below the root zone. Excellent drainage in such situations is mandatory; otherwise, one can expect inhibited seed germination, stunted growth, leaf scorch, wilting and death of the plant.

Mulches reduce water evaporation from the soil. This in turn reduces the concentration of surface salts that result from the upward movement (capillary action) of the soil solution. Mulches thus prevent salt contamination of plant foliages and stems.

Salt accumulation around seeds is a common problem with furrow-planted crops. Instead of seeding in the furrow trough, plant on furrow shoulders, on sloped beds or in two rows on wide beds separated by furrows, so that the salts are carried below seed levels.

If irrigations of salty soils are too infrequent, salt concentrations may reach dangerously high levels between waterings. It is during the periods of highest concentrations that plant injury occurs. Growth and yield will be better if waterings are frequent. A saline soil cannot be allowed to become as dry between waterings as a non-saline soil.

In borderline soils, gardeners may prefer to select crops that are salt-tolerant. Vegetables with the greatest salt tolerance are beets, kale, asparagus and spinach; those with the least are radishes, celery and green beans. Some ornamental plants highly tolerant of heavy salt concentrations are iceplant, coyote bush, bougainvillea, pampas grass and Natal plum; extremely sensitive plants are abelia, rose, Oregon-grape, photinia, cotoneaster and Chinese holly.

Hardpans and Sodic Soils

Sometimes arid soils form a hardpan of lime that inhibits drainage and reduces oxygen in the root zone and drastically raises pH in the pan area. In small garden plots, drainage can be improved by using raised beds or by drilling holes through the layer. Perforating the pan is essential for permanent plantings; without it, each planting hole becomes a basin that holds water and drowns the plant.

Sodic (high sodium) soils are usually impermeable to water. Sodium in soil can be reduced by growing sordon or by applying gypsum. Sordon is a sorghum-sudangrass hybrid with the ability to remove sodium from the soil. Growing this grass is a more economical method of correcting the problem than applying gypsum. Gypsum reacts with other minerals to make soil more permeable and allow it to leach out sodium. To apply gypsum, incorporate it at the rate of five lbs. per 100 sq. ft. Cultivate thoroughly, then water in well. If drainage is only slightly improved, repeat the process.

HOW TO MANAGE SUBSOIL MATERIAL WHEN IT IS AT THE SURFACE

W. LEE DANIELS

Subsoil materials are frequently encountered at the surface of the ground as a result of erosion of the native topsoil or severe soil disturbances associated with earthmoving and construction activities. Unlike topsoil, this material is often quite clayey and dense, devoid of organic matter and resistant to plant growth. Subsoils in humid regions are usually highly leached, acidic and infertile, while subsoils from drier areas tend to be much less acidic, but may be high in shrink-swell clays and/or salts. Subsoils in any region may be rocky.

MAKING TOPSOIL OUT OF SUBSOIL

Usually, the most important factor to correct immediately is the low organic matter content. Large amounts of compost or other organic materials must be repeatedly mixed in *deeply* (one foot or more, if possible). Over time, the organic matter decomposes and stabilizes the new surface soil, aiding in essential soil particle aggregation and building nutrient supplies.

Next, problems of acidity and infertility must be solved through liming and use of appropriate fertilization strategies. Have pH and fertility tests done and follow recommendations. Remember that the establishment and maintenance of organic matter in the soil does much to aid fertility.

Most subsoils are dense and/or clayey, so particular attention must be paid to the problems of poor drainage and water saturation. Even the addition of trucked-in topsoil usually will not solve poor drainage problems caused by heavy or compacted subsoils. Before new topsoil is added, or created by the addition of organic matter, poorly drained exposed subsoils must be deeply, but coarsely, tilled. In some cases laying tiles below grade to facilitate drainage is advised. In many situations the use of raised beds greatly eases the modification of surface soil properties.

Subsoil materials left by construction and earthmoving are often unnaturally mixed and are variable from point to point. Whenever possible when undertaking soil testing, analyze samples from numerous points, rather than putting all samples together; this allows you to isolate the problem areas that need special attention.

COMPACTED ZONES IN SOIL

W. Lee Daniels

Compacted zones or layers in soil seriously limit root, water and air movement and thereby adversely affect plant growth. Soils become compacted by a number of natural and man-made mechanisms, all of which lead to a reduction in soil porosity and an increase in soil bulk density (bulk density is the mass of a dry soil per volume). Soils with low bulk density are desirable because they have more pore space for water and air movement.

A soil which is heavily compacted will severely limit plant growth even if other physical and chemical characteristics such as texture and pH are optimal. Water percolation down through soils is greatly slowed by compaction (or any sudden or major change in soil texture). A layer of compacted soil may cause droughty conditions if at or near a soil's surface, or water logging if the compacted zone is deeper.

Causes of Compaction

Pan is the word soil scientists use to describe compacted soil zones. Natural soil-forming processes generate a wide variety of compacted subsurface pans in all regions of the U.S. In humid areas rainwater leaches clay particles downward into the subsoil, forming clay pans (known to soil scientists as *argillic horizons*). These clay layers are often dense and stiff, making them difficult to till. In fact, many agricultural soils in the East have experienced enough long-term erosion that the subsurface clay pans are now at or near the surface. Plant roots can penetrate these zones along natural cracks and fissures only if the pans have a well developed structure.

A number of commonly occurring compacted zones are collectively known as *fragipans*, and are characteristically very dense and brittle, with few structural gaps for root penetration. These layers are seldom clayey, but tend to be loamy or silty in texture. In drier regions of the country calcium and sodium salts tend to accumulate in the subsoil and often cement it into a dense, whitish-colored pan. If you are unsure about whether or not your soil contains one of these various pans, seek out and study local soil surveys (available from the Soil Conservation Service — your County Agricultural Extension Agent can direct you to the nearest SCS office).

Vehicle traffic, such as associated with construction, can severely compact soils. But normal foot traffic, game playing, even lawn mowing can cause compaction. These compacted zones may occur at the surface or deep in the subsoil, but are often more dense than many natural pans. Artificially induced pans are particularly common where several layers of soil have been disturbed, such as when topsoil is returned to a regraded lawn after house construction, or where cut-and-fill operations have been performed to shape an area for landscaping. Natural soil structure is usually destroyed by these activities; not only are soils made abnormally dense, but there are no longer any natural channels or planes of weakness for roots to penetrate. And naturally occurring pans are, of course, encouraged or compounded by mechanical interference. Soil can even be compacted by rainfall or irrigation, primary causes for the condition known as crusting. Crusting (the formation of a thin layer of surface soil that hardens and resists water penetration) is easily prevented by using a mulch.

Water Impediment Problems

When large amounts of percolating water encounter a compacted zone, or a zone of strongly contrasting soil texture (such as sand over clay or vice versa), water will back up at the contact and saturate the zone above it. Saturated conditions within the rooting zone cause a number of problems for plant growth, including lack of oxygen and potential heavy-metal toxicities.

The nature and quantity of porosity—particularly the amount of large, continuous pores and channels—in soil is the primary factor controlling the rate of water movement. When downward-percolating water encounters a highly compacted zone, it backs up, or "perches," simply because of the major reduction in porosity across the contact. Temporarily perched water tables may persist close to the soil surface from several days to months, depending on local soil and climatic conditions. A similar perching occurs when water passes through a coarse-textured soil layer with many large pores, and then encounters a finer-textured soil layer (even if uncompacted) with much smaller pores. Perching also occurs, but for an altogether different reason, when water passing through a fine-textured layer encounters a coarser sand or gravel stratum. In this case the finer-textured clay soil actually holds onto its water so tightly (due to such forces as capillary action) that it significantly slows its movement into the coarser material below.

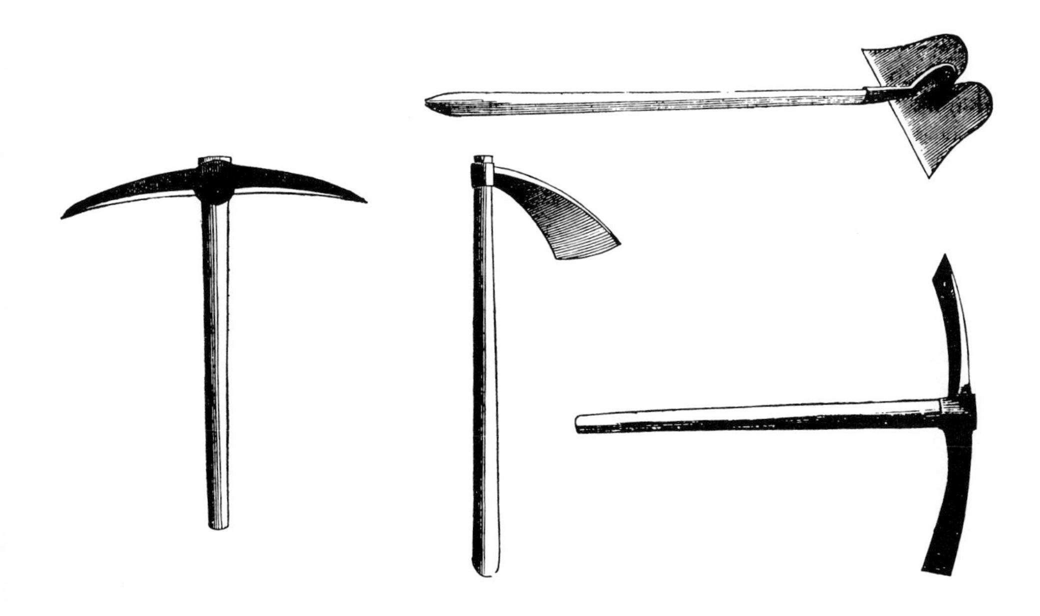

No matter what kind of fertilizer you use, read the label carefully and apply it only at the recommended rates. PHOTO BY ELVIN MCDONALD

ALL ABOUT FERTILIZERS

MICHAEL L. AGNEW

A fertilizer is defined as a material applied to soil to supplement plant nutrients already present. Fertilizer additions to soil are used to increase growth or yield, quality and nutritive value of plants. It is important to have a good understanding of different types of fertilizers when developing a fertilizer schedule for lawns and gardens.

MICHAEL L. AGNEW *is Assistant Professor and Extension Turfgrass Specialist at Iowa State University, Ames, Iowa.*

There are at least three numbers printed on the label of a bag of fertilizer. These figures determine the fertilizer grade and are called the *"N P K numbers"*. The first number refers to the minimum percentage of nitrogen (N) in available form, the second to the minimum percentage of phosphorus (P) as phosphoric acid (P(2)O(5)) and the third to the minimum percentage of potassium (K) as potash (K(2)O). If other nutrients are added into a fertilizer mix, then their minimum percentages are also listed on the bag.

An example of a common fertilizer grade is 10:20:20. A 10:20:20 fertilizer mix has 10 percent N, 20 percent P and 20 percent K guaranteed in any bulk amount. Thus, a 40 pound bag of 10:20:20 contains four pounds of nitrogen, eight pounds of phosphorus and eight pounds of potassium.

Fertilizers are broadly classified as complete, balanced or single-nutrient carriers. A *complete* fertilizer contains all three major nutrients (N, P, K). Examples include 13:13:13, 10:20:10, and 24:4:8. The 13:13:13 formulation is also a *balanced* fertilizer because it contains equal proportions of the three major nutrients. When a fertilizer contains only one of the three major nutrients, it is called a *single-nutrient* carrier. Examples include ammonium sulfate, 21:0:0 (supplies only nitrogen); superphosphate, 0:20:0 (phosphorus); and potassium sulfate, 0:0:50 (potassium).

Selecting a Fertilizer

A factor in the selection of a fertilizer is its acidifying affect. In general, fertilizers that contain the ammonium form of nitrogen produce an acidic reaction, while fertilizers that contain nitrates produce an alkaline reaction. Neither phosphorus nor potassium sources have an important effect on soil pH reaction. Highly acidifying fertilizers should be avoided in regions of the country where basic salts are leached away (i.e., areas with very acid soil); additions of acidifying fertilizers only further increase lime requirements. In areas where the soil pH tends to be alkaline, acidifying fertilizers can help to reduce the pH to a more favorable range. A list of acidifying fertilizers appears in Table 1.

Table 1
Fertilizer Effects on Soil Reaction.

Fertilizer Source	Soil Reaction
Ammonium nitrate	Acidic
Ammonium sulfate	Acidic
Calcium nitrate	Basic
Urea	Acidic
Monoammonium phosphate	Acidic
Potassium nitrate	Basic

Fast Versus Slow Release

The nitrogen fertilizer sour.0ces mentioned up to this point are all *fast-release* or water soluble. While nitrogen in fast-release fertilizers is readily available for plant uptake, its water soluble properties lead to rapid depletion by leaching or volatilization (vaporization into the atmosphere). Also, fast-release fertilizers have a high potential for foliar damage (leaf burn) due to a high salt content of the fertilizer source.

Other nitrogen fertilizers are classified as *slow-release*. These types of slow release nitrogen fertilizers are: natural organics, slowly soluble and coated materials. The release of nutrients from these different materials is primarily due to microbial decomposition, or microbial decomposition in combination with various chemical or physical soil reactions. Nitrogen availability from slow-release fertilizers varies with the time of the year and weather because the activity of soil microorganisms and rate and quantity of soil reaction is dependent on soil temperatures and soil moisture content. The major fast- and slow-release fertilizer sources are listed in Table 2.

TABLE 2
CHARACTERISTICS OF FERTILIZER SOURCES

Source	Approx. N	Nutrient P_2O_5	Percents K_2O	Burn Potential	Availability Rate
Ammonium nitrate	33	0	0	High	Fast
Ammonium sulfate	21	0	0	High	Fast
Urea	45	0	0	High	Fast
Urea formaldehyde (UF)	38	0	0	Low	Slow
Isobutylidene diurea (IBDU)	31	0	0	Low	Slow
Sulfur coated urea	32	0	0	Low	Slow
Plastic encapsulated	V*	V	V	Low	Slow
Activated sewage sludge	6	4	0	Low	Slow
Bone meal	5	27	0	Low	Slow
Cottonseed meal	7	3	2	Low	Slow
Monoammonium phosphate	11	48	0	High	Fast
Diammonium phosphate	20	50	0	High	Fast
Superphosphate	0	20	0	Low	Medium
Treble superphosphate	0	45	0	Low	Medium
Muriate of potash	0	0	60	High	Fast
Sulfate of potash	0	0	50	Low	Fast
Potassium nitrate	13	0	44	High	Fast

*V = amount varies depending upon use.

Natural organic fertilizer carriers include various sewage sludges and plant and animal residues (composts and composted manures). Nutrient release from natural organics is primarily due to microbial decomposition; therefore, nitrogen availability in cool or dry months is low.

Natural organics have the advantages of low foliar burn potential, little nitrogen loss due to leaching and may contain some micronutrients; but unlike processed fertilizers, nutrient content in proportion to bulk material is usually quite low.

Slowly soluble nitrogen fertilizer carriers include ureaformaldehyde (UF) and isobutylidene diurea (IDBU). While UF reacts similarly to natural organics, IBDU is less temperature dependent, but is dependent on soil water content. Thus, IBDU can be utilized during cool weather provided that the soil moisture level is adequate.

Sulfur-coated urea and plastic-encapsulated nitrogen fertilizers are examples of coated slow-release fertilizers. Coated materials are made by encapsulating a soluble fertilizer carrier in a coating of either sulfur or plastic. Nutrient release occurs following the gradual diffusion of water into the capsules. The release rate of the fertilizer is controlled by the thickness of the coating. The nutrient release of the coated materials is only dependent upon amounts of soil moisture, so coated materials release nutrients uniformly during periods of cool or warm weather as long as moisture levels are uniform.

Forms of Processed Fertilizer

Fertilizers may be mined or manufactured, and are formulated as granular, liquid or spikes. Granular fertilizers are the most common. The physical characteristics of granular fertilizers are set by mechanical screening so that spreadability is easy and accurate. Liquid fertilizers are sold either as liquid concentrates or as dry base materials; either must be diluted or dissolved in water prior to application. Both granular and liquid fertilizers can be used for all types of plants (trees, shrubs, flowers, vegetables, fruits and turfgrasses).

Tree and shrub spikes are examples of fertilizers formulated for special purposes. A major concern about spikes is the high concentration of fertilizer in one spot. There is no guarantee that the tree and shrub roots will be in that one location. In addition, the highly localized concentration of fertilizer may burn a root that grows into the area. For this reason, make sure you use spikes that contain a slow-release source of nitrogen.

Types of Plants and Fertilizers

The grade of fertilizer recommended for a particular plant group varies depending on what the plant or crop is grown for. For example, a turfgrass fertilizer is normally complete, with a ratio of 3:1:2 or 4:1:2, and containing at least 35 percent of the total nitrogen as a slow-release material. This is also a good formulation for most trees and shrubs. In contrast, a vegetable fertilizer varies between leafy crops, root crops and pod or fruit crops. Leafy crops require large amounts of nitrogen, so a 1:1:1 or a 2:1:1 fertilizer ratio is recommended. But a root crop requires more potassium; a 1:1:2 or 1:2:2 fertilizer ratio is best. In the case of fruit-bearing vegetables such as tomatoes or peas, more phosphorus is required, so a 1:2:1 or 1:2:2 is the preferred ratio. This rate is fine for flower gardens as well. For vegetables most nitrogen is supplied by a fast-release source.

SOME PLANT GROWTH FUNCTIONS OF SOIL DERIVED NUTRIENTS

Nitrogen	A constituent of all proteins and nucleic acids (hence all protoplasm).
Phosphorus	Fundamental role in many enzymatic reactions, essential to cell division and the development of meristems (growth points).
Potassium	Important to the synthesis of amino acids and proteins, also necessary for photosynthesis.
Calcium	Essential to the growth of meristems, particularly root tips.
Magnesium	A constituent of cholorophyll, necessary for the transport of phosphorus in the plant.
Sulfur	Constituent of many proteins and essential oils.
Chlorine	(Chlorides) Osmotic pressure regulator and a cation (ion) balancer in cell sap (very rarely deficient).

Note: Many deficiencies in plants are due to a particular species' special needs.

Whether a fast-release or slow-release fertilizer is used depends on plants treated, application schedules and personal preference. Either does the job if applied properly. Care must be taken not to overapply fast-release sources of fertilizer, or fertilizer material is wasted and plants may be damaged. Conversely, slow-release materials should not be underapplied, especially the synthetic and natural organic sources, or plant nutrient deficiencies may occur. ❦

WHAT IS FERTILIZER BURN?

Fertilizer burn is the desiccation (drying out), or necrosis (browning and death), of plant tissue caused by improper or overapplication of fertilizers.

Fertilizer burn is caused in three ways: 1) by the physical contact of concentrated fertilizer particles or very fresh manure with plant parts; 2) by the volatilization (gas formation) of fresh-manure ammonia that may accumulate around growing plants or plant roots; 3) (and this is the most significant) by an overabundance of nutrient ions or salts in the soil. In this case the "burn" is caused by drought: because of the specialized way that plants take up water, if the ion concentration in the soil solution reaches above a certain level, then water can no longer enter plant roots. The concentration of ions in soil water must be less than the concentration of ions in water in the plant.

Fertilizer burn is affected by humidity, amount of soil moisture, rainfall or irrigation, wind and air temperature. Each of the three types of burn is affected differently by these factors.

To avoid fertilizer burn there are reliable rules-of-thumb to follow: never let processed fertilizer particles remain on plant leaves; use fresh manure sparingly or only in composted form and apply fertilizer only at recommended rates (two lbs. or less of nitrogen per 1,000 square feet. per year is a safe rate of application for most plants, soils and situations).

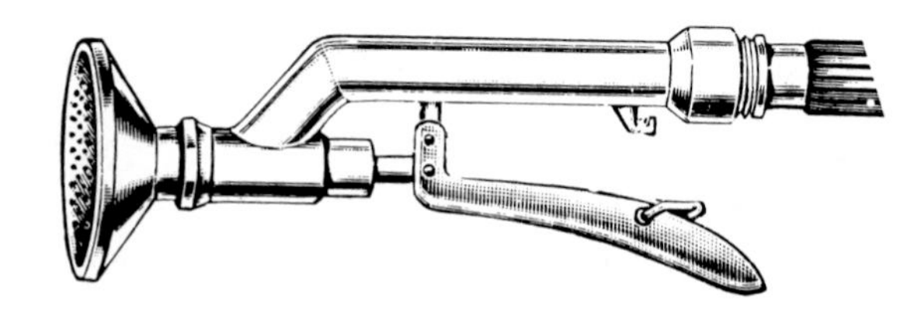

WHAT IS FERTILIZER? WHAT IS COMPOST? IS THERE A DIFFERENCE?

Which is Best?

Fertilizer is any material used to supply nutrients to plants. Compost is any organic material that has mostly or fully decomposed. Decomposition is a process that breaks larger compounds into smaller ones. Since manure should be used only after it is decomposed somewhat, manure you use in your garden is also compost. Fresh manure and just-begun composts are undesirable because they either contain toxic compounds or have not broken down enough to release nutrients in forms plants can use. But in general, composted manure is called just manure—sheep, cow or whatever to indicate its difference from plant-derived or blended (manure and plant parts together) compost. Usually, compost is referred to as a soil conditioner or as a material used to fertilize (i.e., "I'm going to fertilize with compost"), rather than as a "fertilizer."

Processed fertilizer, manufactured or refined products that are sold in bags and boxes, is what is usually thought of or referred to as "fertilizer." Processed fertilizer can also be composted manure or some other compost, but it is usually composed of manufactured ingredients and/or compounds derived from natural sources such as guano (bird droppings) or very pure rock minerals. Processed fertilizer has an enormous advantage over compost because its content is standardized. A farmer or knowledgeable gardener can thereby supply exactly the kind and amount of nutrient needed. Processed fertilizer is also much more concentrated. For example, 50 lbs. of 20:20:20 may equal a ton of cow manure in total quantity of available nitrogen. Although composts and manures vary in nutrient content (because of plants used or eaten), they usually have a nitrogen content, by weight, of between .5 and 10 percent (if figuring application rates for fertilizing purposes, a 2 percent nitrogen content is a safe guess for most composts and composted manures). On the other hand, composts have an advantage over processed fertilizer in that their fiber content helps improve soil structure. Also, unlike processed fertilizer, composts and manures contain essential elements other than nitrogen, phosphorus and potassium. But completely decomposed composts and manures often lose their effectiveness because much, sometimes all, of their useful nutrients are leached away.

Organic has two meanings: 1) any compound containing carbon; 2) any material derived from a living source. The first definition is used when discussing chemicals and biological compounds, the second when referring to the remains or by-products of living things.

The bulk of the nutrients that plants get from soil—nitrogen, potassium, and phosphorus—are inorganic (do not contain carbon). Nutrients are identical whether derived from an organic source (such as manure) or an inorganic source (such as processed fertilizer).

So which is best to use—processed fertilizer or compost/manure? That's a decision for you to make depending on availability of materials, desired soil management, time and funds.

SOME NUTRIENT DEFICIENCY SYMPTOMS

NITROGEN	Small leaves; short, small, thin growth. Uniform yellowing of leaves, becoming red in severe cases. Leaf abscission (removal) occurs beginning with older leaves.
PHOSPHORUS	Bronze to red coloration of dull-green leaves. Shoots short and thin.
POTASSIUM	On older leaves, marginal chlorosis (yellowing) followed by marginal necrosis (death or dead areas).
CALCIUM	Death of the terminal bud, often preceded by chlorosis of the young foliage. Leaves may be distorted, with the tips hooked back. Root system is damaged first, usually with death of the root tips.
MAGNESIUM	Interveinal (between leaf veins) chlorosis on the foliage, followed by interveinal necrosis.
BORON	Distortion or thickening, or both, of the young terminal foliage. Some yellow spots may appear on the foliage. Terminal bud ceases development and lateral growth begins. Lateral growth soon goes through same cycle.
SULFUR	Uniform yellowing of new and young foliage.
IRON	Interveinal cholorosis of the young foliage, followed by bleaching of leaf color to cream or white.
COPPER	Necrosis and white mottling of newer foliage. May result in small, linear, distorted leaves. Shoots die back.
MANGANESE	Yellowing of foliage.
MOLYBDENUM	Leaves do not fully expand; color often bluish-green; chlorosis and necrosis also present.
ZINC	Small, narrow leaves in a rosettelike whorl.

Because nutrient deficiency symptoms of different elements are sometimes similar, because the same nutrient deficiency on different plants may be manifested differently and because some diseases cause similar effects, positive diagnosis of nutrient deficiencies is very difficult. Although with experience some deficiencies, especially those that are nitrogen related, are fairly easy to identify, the best means is through soil tests or plant tissue analysis. Many states offer plant tissue analysis through the State Extension Service.

Nutrients, and other minerals, can be overabundant in soil and cause toxicity. This overabundance can be natural or caused by man (chlorine from pool water is an example). Natural soil toxicities are fairly uncommon, and their symptoms and signs, unfortunately, often resemble mineral deficiencies, diseases or such things as air pollution damage (which of course is a toxicity, too).

TOSS IT OUT
OR MEASURE CAREFULLY

A GARDENER'S FERTILIZATION DILEMMA

Fertilization recommendations are much more specific these days than they once were. It used to be that you were simply told to put a handful, or 1/2 cup, at the base of a shrub or scatter fertilizer lightly under a tree's canopy.

But now it seems that a weight scale is more important than a measuring cup. However, the measuring cup is more convenient—so what do you do?

First of all, do not guess. Fertilizers come in a wide range of concentrations and 1/2 cup of one may be just what the doctor ordered, while 1/2 cup of another may send a plant to the coroner. Do learn to figure amounts of nitrogen per 1,000 square feet, and at least the first time you use a product for a specific plant or plant group, do measure it out. After that you can probably safely "guestimate" amounts to be applied.

However, large areas requiring significant amounts of fertilizer such as lawns and vegetable gardens should never be figured "by the seat of your pants."

Fertilization is a tool that helps you grow plants bigger, better and faster. Use it well and you'll get favorable results; use it poorly and you'll waste the product—or the plant.

METHODS OF FERTILIZING

Surface Application, the Best Way

Mary L. Wilmoth

Neither soil injection nor the placement of pills, packets or spikes surpasses broadcast surface application in fertilizing effectiveness. Surface application is the easiest and fastest way to apply fertilizer because little or no equipment is necessary and the labor involved is minimal. And research has shown that in most cases, surface applications for established plants are as effective, or more effective, than subsurface treatments. This is probably due to the localizing effect of drill-hole type procedures, as opposed to the more even coverage of broadcast applications. For proper surface application, spread the recommended amount of fertilizer *uniformly* on the ground under and beyond the branch spread of treated plants.

Recommendations of over three pounds of actual nitrogen per 1,000 sq. ft. per year is too high for a single application around trees that have grass or other plants growing under them. For such situations it is necessary to apply fertilizer in several small applications (with thorough watering after each).

Mary L. Wilmoth *is Propagator/Horticultural Assistant at The Dawes Arboretum, Newark, Ohio.*

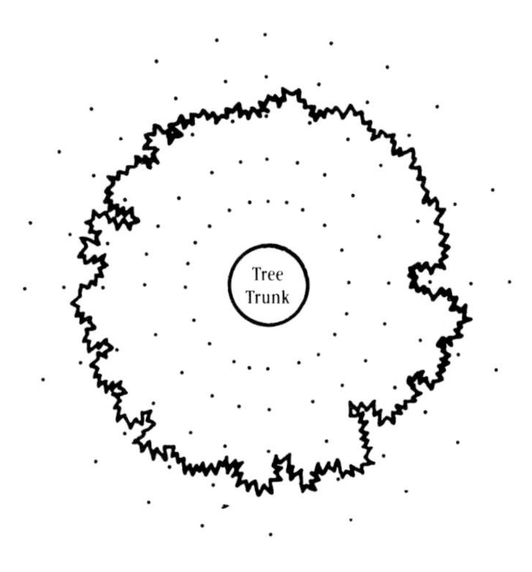

Recommended pattern for subsurface applications of fertilizer: drill-hole or liquid soil-injection methods.

Limited-Use Methods for Homeowners

Liquid Soil Injection. This involves the use of a thin tube pointed and perforated at the end. It is attached to a hose or sprayer, and fertilizer solution is injected through the tube and into the ground at the desired depth. Insert the tube to a depth of six to 12 inches on

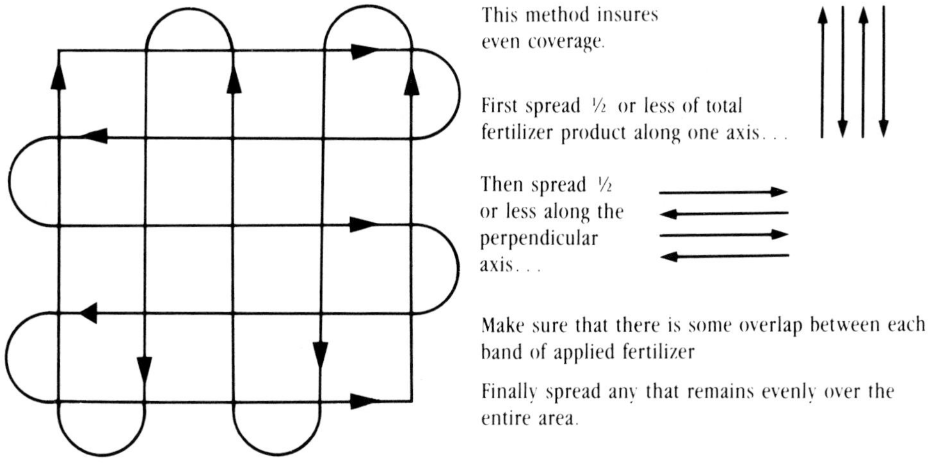

RECOMMENDED PATTERN FOR SURFACE APPLICATIONS OF FERTILIZER

approximately two-foot centers. Liquid soil injection is a relatively rapid means fertilizing below the surface of the soil and turf roots and irrigating at the same time. Do-it-yourself devices that attach to a hose are not effective for applying fertilizer because the amount applied is usually far below recommended levels.

DRILL-HOLE APPLICATION. Punching or drilling holes in the soil and then placing fertilizer in them is a technique that has the advantage of aerating the soil, as well as feeding the zone below turf roots. It also permits the addition of organic or other materials that improve soil conditions. This method is especially beneficial if plants are growing in compacted or poorly aerated soils; the improved aeration may be as effective as the fertilizing in stimulating new root growth. One disadvantage is the amount of labor required to make many holes necessary for even distribution. A motor-driven auger or a punch bar can be used, although the auger is preferred. Augers can be obtained at tool rental centers. The job is done most easily when the soil is moist but not wet.

Dig holes on 12- to 18-inch centers to a depth of six to 12 inches, beginning two feet from the trunk and extending in concentric circles to a line beyond the edge of the branches. If fertilizer in a hole is within two to three inches of the surface it can either kill grass or stimulate a prominent clump of lush, green growth. Because most fertilizer elements do not move laterally in the soil, the more holes made, the better the distribution throughout the root zone. The recommended amount of fertilizer should be distributed evenly among the holes, using a funnel and a small cup as a measuring device. For best results, fertilizer should be mixed with a carrying or filler material — soil, peat moss or sand — so that it won't be too concentrated. After fertilization, completely refill the holes with soil.

CALCULATING FERTILIZER APPLICATIONS

Application-rate recommendations are usually in pounds of nitrogen (N) per 1,000 sq. ft. (but any area division can be used). Chemical analysis numbers, such as 10 in 10:10:10, refer to pounds of nutrient per *100 pounds* of fertilizer. The first number of the three *always* represents nitrogen.

1. To figure the number of pounds of fertilizer that supply one pound of nitrogen:
   EXAMPLE: For a fertilizer rated 10:10:10 (lbs. of nutrients per 100 lbs. of product), divide 100 by 10 (100/10 = 10), or 10 lbs. of fertilizer product contain one pound of nitrogen.

2. To figure the number of pounds of fertilizer that equals other than one pound of actual nitrogen:
   Figure for one pound of actual nitrogen, then multiply answer by number of pounds desired. Example: Recommendations call for two pounds of actual nitrogen and you want to use 18:5:9. Remember, the figure "18" refers to the pounds of nitrogen per 100 lbs. of total fertilizer. Divide 100 by 18 and multiply the answer by two (100/18 = 5.55 x 2 = 11.1). Or 11.1 lbs. of fertilizer contains two pounds of actual nitrogen.

3. To figure the amount of fertilizer to apply for areas greater or smaller than 1,000 sq. ft.:

   a) Figure the number of square feet and then the number of pounds of nitrogen. Example: A 135-by 70-ft. lawn = 9,450 sq. ft. 9,450 sq. ft. = 9,450 increments of 1,000 sq. ft. If 1.5 lbs. of nitrogen per 1,000 sq. ft. is recommended, then 9.450 x 1.5 = 14.175 lbs. of nitrogen is needed to apply 1.5 lbs. of nitrogen/1,000 sq. ft. to 9,450 sq. ft.

   b) Figure the number of pounds of fertilizer that equals one pound of nitrogen (see No. 1). Example: If fertilizer used is 24:8:12, then 100 divided by 24 = 4.17 lbs. contains one pound of nitrogen.

   c) Figure the total number of pounds of fertilizer product required. Example: The pounds (rounded off) of nitrogen needed to meet recommendations (14.18) is multiplied by 4.17 (the pounds of fertilizer containing one pound of nitrogen) to get, approximately, 59 (which equals the pounds of fertilizer to be spread over a 135- by 70-ft. lawn to achieve an application rate of 1.5 lbs. of actual nitrogen per 1,000 sq. ft.).

LEAD IN SOILS

David C. Martens and Grover G. Payne

Because of its physical and chemical properties, as well as its relative abundance, lead (Pb) has been an important metal in human society for thousands of years. Currently, more than 2.5 million tons of lead are produced each year from mined ores. This lead is used in the manufacture of a number of common products including storage batteries, gasoline and paints. But lead poses a potential health hazard to living organisms. The accumulation of lead in the environment has been the subject of increasing concern in recent years.

Lead occurs naturally in all soils, with levels in unpolluted soils generally ranging from two to 200 ppm depending on the nature of the soil's parent material. Additional lead-containing substances are commonly introduced into the environment from a variety of sources. The most common include automobile exhaust, industrial emissions from smelting and refining operations, and the use of sewage sludges as soil amendments.

Exhaust from automobiles is thought to be the primary source of lead pollution in urban areas. Many studies have shown that elevated levels of lead exist in the soil and plants adjacent to highways. Concentrations depend on the traffic volume in the area and the distance from the road. Soil lead content in many major metropolitan areas of thee United States is reported to be several times higher than in rural areas due to the accumulation of lead from automobile emissions. But in most cases soil lead concentration decreases rapidly with distance from the roadway. This fact should be taken into account when choosing a garden site.

Increasing amounts of lead-rich municipal waste materials are being applied to soils as amendments or as sources of plant nutrients.

Repeated applications of these materials can cause substantial increases in the

David C. Martens *is a faculty member in the Department of Agronomy, Virginia Polytechnic Institute and State University. He has a Ph.D. in soil science.*

Grover G. Payne *is currently completing requirements for his Ph.D. in soil science at Virginia Polytechnic Institute and State University.*

lead content of soils, eventually reaching levels that are harmful to animals *and* plants if proper precautions are not taken. Even when lead concentrations are not high enough to be toxic to plants, they may be to us if the plants are food crops. We must *never* raise edible plant materials in lead-contaminated soil.

Emissions from lead smelting or refining operations are generally considered localized phenomena, but they can cause severe problems in the environment immediately surrounding the source of emission.

Another location where elevated lead levels are often found is in the soil around houses (particularly older ones) that have been painted with lead-based paints. This source of environmental lead contamination is not generally considered as serious as others described, except in cases where paint chips are ingested directly by humans or animals.

Once in the soil, lead is relatively immobile and unavailable for plant absorption because it bonds strongly to various soil components. Lead forms very stable insoluble compounds with phosphorus and humic (organic) substances in the soil and becomes strongly held by clay and oxide minerals. Another important factor in controlling the availability of lead is the pH of the soil. At high soil pH levels, lead forms insoluble hydroxide compounds which are not taken up by plants. As the soil becomes more acidic the solubility of these compounds increases and more lead is available for plant uptake. Therefore, to minimize plant absorption of lead from contaminated soils, it is essential that the pH of the soil be maintained at 6.5 or above. Further steps that can be taken include maintaining generous organic matter contact and phosphorus fertility status in the soil.

Soil concentrations of lead must reach relatively high levels before the amount taken up by plants is increased significantly and detrimental effects to the plant are observed. Most plants accumulate lead in their root systems, with only limited amounts being translocated to the above-ground portions of the plant. The part of the plant harvested is then an important consideration in selecting the types of plants grown in lead-contaminated soils.

Because of the immobility of lead in soil and in the roots of plants, the major portion of lead found in the aerial parts of plants is thought to come from direct contamination of the foliage by air-borne lead substances. Since most air-borne lead pollutants settle to the ground within short distances from where emissions originate, it is advisable to locate gardens as far away as possible from potential sources of contamination.

If you suspect high levels of lead in soil around your home or in your garden, take a soil sample and send it off for testing; be sure to indicate that you want the sample analyzed for lead content.

PHOTO BY ELVIN MCDONALD